ISRAEL KLABIN

Urges of the Present

FOREWARD
FERNANDO HENRIQUE CARDOSO

ISRAEL KLABIN

Urges of the Present

A BIOGRAPHY OF THE ENVIRONMENTAL CRISIS

VERSÃO RENATO REZENDE

ELSEVIER

CAMPUS

Cover drawing: Vicente Klabin Salles
Translation into English: Renato Rezende
Copydesk: Vicky Adler
Desktop Publishing: Estúdio Castellani

Elsevier Editora Ltda.
Conhecimento sem Fronteiras
Rua Sete de Setembro, 111 – 16º andar
20050-006 – Centro – Rio de Janeiro – RJ – Brazil

Rua Quintana, 753 – 8º andar
04569-011 – Brooklin – São Paulo – SP – Brazil

Customer Service
0800-0265340
sac@elsevier.com.br

ISBN 978-85-352-6156-1

CIP-Brasil. Catalogação-na-fonte
Sindicato Nacional dos Editores de Livros, RJ

K69u Klabin, Israel
Urges of the present: a biography of the environmental crisis / Israel Klabin ; [tradução de Renato Rezende]. – Rio de Janeiro : Elsevier, 2012.
21 cm

Tradução de: A urgência do presente : biografia da crise ambiental
ISBN 978-85-352-6156-1

1. Mudanças climáticas. 2. Mudanças ambientais globais. 3. Meio ambiente. 3. Impacto ambiental. 4. Desenvolvimento sustentável. I. Título.

12-2419. CDD: 363.7
CDU: 502.1

To future generations

I have the habit and professional duty not only at home, but also at work, to be around young people wanting to learn something from me. This has been the very reason why an aura of youth is transferred to me and allows me to refer to and put into practice what I have learned from my elders.

Cover drawing: Vicente Klabin Salles, 4 years old

Foreword

This fascinating book strikes the reader, starting with its unusual title expressing a sort of riddle about humanity, gradually revealed in these very appropriate lines, which could only be written by a person with the intellectual gravity of Israel Klabin. It is not, strictly speaking, a biography, but it shows the journey of the idealist man who left business management to go in search of knowing "what was beyond the bend in the road." The business world lost a professional talent, but environmentalism won a new fighter. These pages captivate those who love deeper reflections about men and nature.

Urges of the Present surprises for several reasons. Its prose is well written and enjoyable to read. Filled with information, it reveals crucial details of historical events within the journey of global environmentalism, from the Stockholm Conference to the recent Conferences of the Parties (COPs) on climate change and biodiversity. Nothing escapes the descriptive impetus of the author, who pays attention to all themes in the environmental agenda, from the biomes to the Amazonian "flying rivers," from the energy matrix to carbon sequestration, from China to Africa, from green economy to ethics, from corporate governance to responsible government. Israel demonstrates incredible knowledge with a modern approach, without ideological biases or intellectual prejudices, drawing the lay person to share in the dilemmas of the global ecological crisis.

Israel makes it clear that, from the point of view of the architecture of the new economy, "we are making the wrong calculations" when we consider GDP wealth, neglecting the depletion of natural resources and global warming. As a businessman, he talks about responsibilities and huge opportunities, and about the private sector in the economy of the future.

The author highlights the protagonists of his time, especially people who have contributed to the construction of a global ecological thinking. Thus he lets us know, for example, that the famous report *The Limits of Growth,* by the Club of Rome, was led by Dennis Meadows and was kicked off by the Italian Aurélio Peccei, president of FIAT and one of the founders of the Club. I could highlight other equally revealing passages, some even including myself. Instead of trying to steal his scene, as is often common, Israel does not promote himself, but focuses on the social and political change needed in the path to a sustainable world, chronicling the contributions of many characters. He even fills the texts with philosophical and poetic quotations, lending brilliance to his argument.

I met Israel Klabin a long time ago. A man with a long and diverse background (member of the Brazil / U.S. Commission of the Alliance for Progress; founder of ISEB with Roberto Campos; active in the creation of SUDENE (1959); Mayor of Rio de Janeiro in 1979 and so on), like any active participant of the Brazilian scene I was very likely to meet him at some point. He was a friend of Mário Henrique Simonsen, Helio Jaguaribe, San Tiago Dantas, Octávio Bulhões and of so many others that it would be impossible for any prominent intellectual or politician to be his contemporary and not know him.

Israel Klabin's prestige is so great and spread out that once in Lagos, Nigeria, I gave a conference that was attended by another remarkable character, the Ashanti King. He attended my lecture dressed in ceremonial costumes. He solemnly entered the room with a half naked chest, wearing gold necklaces and bracelets, under a canopy followed by a few acolytes who knocked firmly on the floor with thick sticks while others trumpeted instruments that looked like bull horns. He greeted me and addressed the public with a few sentences in impeccable British English. After the ceremonial part, the first question he made was:

"Do you know my friend Israel Klabin?"

The magic wand uniting Israel to that character was the passion both shared for environmentalism, the same that drew me and him and earned him respect in so many different places and from so many different people.

Not only that: when I was a Senator, I recall a dinner at his house on the hill of Vidigal, São Conrado, which Henry Kissinger – then the powerful chief of U.S. Foreign Affairs – also attended. Israel was treated with respect by the guest and one could understand that their relations were of esteem and independence.

Since he left his business activities and, having participated actively in the UN sponsored conference in Rio de Janeiro, the Earth Summit, he ended up "fertilized by the idea of environmentalism," Israel Klabin dedicated his intelligence and willingness to work in the Brazilian Foundation for Sustainable Development (FBDS), of which he is president since its founding in 1992. In this capacity, he was actively present in the most important meetings in Brazil and abroad on a subject that is truly dear to his heart. An advocate of global actions, he compares nationalism, remembering Kant, to the "crooked timber of humanity."

After the Earth Summit, Israel, already head of FBDS, had a crucial meeting in Kyoto, when he and the representatives of my government worked hard on the definition of clean development mechanisms relating to carbon credits, which would interest emerging countries outside Annex I, such as Brazil. I personally did all I could for developed and developing countries to reach an agreement. I remember that there was a moment in which Clinton called me asking for my interference: the countries of the "South" believed that the industrialized countries were responsible for the pollution and, therefore, only they should pay for the cost of environmental cleanup. In fact, we evolved toward cooperation, but with differentiated responsibilities, leaving the heaviest burden to the developed countries. The Kyoto Protocol was an important step in this regard, requiring far more energetic steps.

Watching Israel talking about the water and animals of the Pantanal – especially for those who may have the good fortune I had to be with him in the region – shows that his love of nature goes to the utmost depth of his existence. He writes with real conviction and passion, and above all does it with an optimism that causes envy in Maurice Strong, one of his main interlocutors. His penchant for the universal – a very Jewish tradition — does not dissuade him from looking at the details and viewing them as a concrete expression of broader trends.

Anyway, if a humanism we can relate to is still alive it is because there are people with the gifts of intelligence and generosity needed to pull the best from each one of us in order to benefit of themselves, of others, and of the survival of our species with a long-term vision of sustainability. This book is a testimony that, luckily, there are still those who continue to firmly keep the flame of these lofty ideals burning.

Fernando Henrique Cardoso
São Paulo, September 2011

Acknowledgements

In order to write this book in close collaboration with so many friends, colleagues and various actors in different situations, I was allowed to compose a multidisciplinary work. I would like to thank those who, for one reason or another, I have not mentioned in this first page.

Here is my nostalgic remembrance of those who are gone and who taught me: my wise teachers, those I chose and who enlightened me with their knowledge and work.

However, this book, to which I lend my name, is not only mine, but also of the masons, carpenters, architects and engineers who helped me. I would like to thank Cristina Aragão, a journalist who, above all, is a universalist, and her assistant, the young historian Leonardo O'Reilly Brandão. Sometimes I wonder if I should have included them as co-authors. The patience and dedication they had with this work over more than a year forced me to make an effort of great discipline and to tame my own time. They followed me, looking for gaps in my very busy schedule and, through a secret alliance with my assistant, Débora Wainstock, I was imprisoned in my office where, in between interviews, dictations and recordings, we managed to reach the end of the proposal contained in this book.

The participation of my team from FBDS was of paramount importance. I thank my colleagues for their support: Professor Enéas Salati, Walfredo Schindler, Clarissa Lins, Agenor Mundim and Branca Americano.

The responsibility for the ideas is totally mine. However, the connections to the concepts developed here belong to everyone. I am also grateful to my fellow friend, Fabio Feldmann, who not only encouraged me to write this book, but was also the first critical reader of the completed text.

My greatest gratitude goes to my wife, Lea, and my children Alberto, Maurício (*in memoriam*), Leonardo, Stela, Maria, Dan and Gabriel. They are the people who make me aware of the present and it is because of them that I believe in the future.

CONTENTS

Introduction

WHAT YOU WILL FIND IN THE FOLLOWING PAGES

This book consists of my own questions related to the urgencies of the present moment. For the first time in history, we have an issue common to all mankind. It is not a question of *us* and *them* anymore. We live in a defining moment of transition. There is no time to waste.

Everybody has already heard about climate change. Even though these words do not directly touch hearts and minds, it is more than evident that human actions are what is disrupting the Earth's thermal balance. Climate disasters, accelerated desertification, and severe losses in the biosphere are just some of the consequences.

Climate change is a theme which is subject to many interests, facing enormous resistance within systems of power. There are vested interests, and many people who want the game to fail. It is the opposite of a win-win situation. Climate change is not a cause, but an effect of political and economic models. While these models are not adapted to the urgent needs of this transition, we will not touch the causes.

The Earth is our home. Ecology is a word constructed by two Greek words: *eco* (home) and *logos* (order), thus, we need to organize our home. Mankind has changed the world. Scientists argue that we have caused such drastic changes just in the last two centuries that we are inaugurating a new geological epoch. We left the Holocene, a stable era which started 11,000 years ago, to the Anthropocene, a new geological era caused by human actions. We are transforming the earth very quickly.

Our interference profoundly affects physical, chemical and biological cycles which ensure the existence of life as we know it. By burning fossil fuels, producing fertilizers, or diverting the course of water, we alter important natural flows. Species are being lost daily in one of the greatest mass extinctions that has ever occurred on Earth. In the future, geologists studying our time will conclude that something unprecedented was happening.

Mankind has always reacted to the forces of nature as well as being dependent on them. Technology and industrialization have sought to disrupt this dependence, but today we must recognize that we really are just part of a natural system. Seventy-five percent of greenhouse gas emissions come from the burning of fossil fuels. Science has established two important reference years in our century: 2020 and 2050. Emissions should be stabilized until 2020 and drastically reduced (80%) in 2050 if we do not want the average increase in temperature to exceed two degrees Celsius by the end of the century. This was established in Copenhagen in 2009. Therefore, the present decade is crucial. If this chance is lost, a later attempt will have to be done with shorter deadlines and higher financial and political costs.

We need to understand very well that when we talk about temperature increase on Earth, we are talking about the effect of this increase on the climate. If the current business as usual scenario continues, our lives and the lives of our grandchildren will be compromised. This is not futurology, this is science.

There was a time when developing countries, facing the great powers, felt they had the moral right to increase their emissions of greenhouse gases. There are no rich or poor anymore, because everybody is now facing the same problem. There is no place in the world to appease people anymore. Geographical constraints are no longer limiting factors as they were in the past. These days, climate change is the most comprehensive and urgent issue.

The year of 1992 was seminal in the history of mankind because, firstly, we recognized that we had a common problem. The Earth Summit was also a defining moment in my own life story. At that moment, I was fertilized by the idea of environmentalism, as will be seen in the following pages. In retrospect, I see the Earth Summit — also known as the Rio Summit and Rio Conference — was almost prophetic. The difference today is that the problems identified twenty years ago have intensified, at the same time as the solutions are now clearly within our reach. Why do we not implement them, if we already have a multitude of tools we did not have in 1992? In 2010, CO_2 emissions broke the historical record in energy, transport, and industry.

The question that arises is: why do we not react? When a discourse like this comes into play, all matters become political by definition. What we see is that there is no political will to condense solutions into a common project.

And why the urgency of the present, the expression I chose as title and summary message of this book? Urgency here does not fit its most everyday, ordinary meaning. The world is not on the brink of ending tomorrow, or next year; there is still time to solve our problems. However, if we do not do it in this decade, in 2020 our world will have already become extremely problematic, and in 2050 we may have a completely different Earth than the one we know today. The effects of structural decisions today will be measured in the near future.

It is always useful to look into the past to find references for the present. The biggest meaning of the urgency I refer to can be condensed in the *Triskelion*, or three legs symbol. Present in the culture of ancient Greeks and Celts, the *Triskelion* can be represented by three legs moving symmetrically placed in a circle, facing the outside and united in the centre. An ancient coin with the *Triskelion* symbol was given to me by a friend, Karl Katz, when I was finishing this book.

The three legs, in perpetual circular motion, in our times mean the need to take urgent decisions about the three great structural axes of civilization in the spotlight: the environment, the economic model, and the political model. Our common future will be shaped by these three fundamental dimensions. Politics, economics, and ecology. The three legs complement each other; they are three axes of action which need a fundamental change of perspective. On the coin, the three legs are contained in a circle which, to me, symbolizes the Earth: it is the planet that gives the necessary and vital limits to man, his systems and processes.

Since the Earth Summit, I attended many COPs (Conferences of the Parties), and to what conclusion did I arrive? We speak of the effects of climate change, but not about the causes. We do not touch the real problems. Power systems provide services to business-as-usual structures, like the oil sector, the framework of the world of finance, and multilateral institutions which govern the global macroeconomy.

The domestic issues that both statesmen and their delegates cling on at climate conferences will be lost in irrelevance if environmental issues are not considered a priority.

A new mission was given to me 20 years after the Earth Summit: to help shape the contents for Rio+20, the UN conference on Sustainable Development which will take place in Rio de Janeiro in 2012. It is a historical opportunity we should not waste. During the Earth Summit, common responsibilities were addressed. Now, in 2012, it is important to define responsibilities. It is essential that political conditions for Rio+20 are created. Governance will surely be one of the most controversial issues and cause the most heated debates.

There is no conflict between environmental sustainability and economic growth. Instead, Rio+20 intends to catalyse efforts around green economies. The theme is not only relevant for the developed economies, but also for growth and the eradication of poverty in developing economies.

Thus far, the value of natural capital – the guarantor of life and productive and intelligent presence of man on earth – has not yet been established. The first necessary condition for a new development is to return the respectability of that currency, which should be anchored in renewable or non-renewable natural resources.

Another question that arises: why do we not advance more in implementing clean energies? Because interests in the maintenance of an unregulated economic model do not allow us, as they are closely connected to the political, industrial, and military models.

We are now almost seven billion people on Earth. The rate of man's predation on the environment is far greater than the speed at which natural resources can regenerate. In a likely scenario, we will be two billion more by 2050. This huge population will put further pressure on the natural resources demand.

In the twenty-first century, water will be more precious than oil was in the twentieth. We depend totally on the ecosystems to continue this extraordinary experience called life.

Climate change is part of an open, multidisciplinary system, encompassing both the exact and the human sciences. For this reason, while preparing this book, I found relevant the interweaving of all sciences. Here I raise philosophical, historical, and affective questions.

This book is not a biography. I chose to resort to a magnifying glass – often historical, often experiential – to report the facts that follow. I returned to letters exchanged throughout my life with many companions. I revisited authors, met with my old teachers, and remembered the wise ones. And why not? I brought back poetry. There are moments in which poets are the ones who know best. Nothing is more enlightening than an act of grace!

The writing of this book rekindled many moments and circumstances within me. Drummond observed that "The magic of the old papers keep on living if we know how to read them."[1]

Here, I do not pretend to give easy answers to vital needs. What I want to show is that we environmentalists are not idle dreamers. The fools are those who still believe it is not possible to reconcile development with ethical policies, clean energy, and social inclusion.

But for that to happen – as there is always a *but* in every story – we must improve economic and political models. We are just at the initial stage of what these new models might become. This book is an invitation to reflect, in order to enable us to seek together possible ways to guarantee our descendants the same reserves of natural resources we received from our ancestors.

This book is addressed to all who are concerned with the challenges of the present and future, and is dedicated to those who still cannot read: the children of this and the next generations.

Israel Klabin
September 2011

Preface

I needed to know what was beyond the bend in the road. In my mature age, I sought a new professional destiny.

At that time, no more than a few hundred people considered environmentalism a science, and there were only vague guidelines on the concept of sustainable development. The idea of sustainability was still in its early stages. I had only a few notions that it could be where the path would lead me. But I had enough foresight to know that things could not continue the way they were.

I was going through a sabbatical period; I had decided to leave business management and was rethinking what would be my new direction in the world.

Some phenomena transcend the limits of humans to understand them as simple facts. We have built a world structured by short-term decisions. It was time to try and build in the long term.

Life put me against deadlocks in the relationship of man with his economic and political frameworks, making me sure about how inviable the model we lived with during the twentieth century was, the model with which we built a society. I come from a family that 200 years ago lived off the conservation of boreal forests in Lithuania, the only alternative for those poor peasants. So I have had an umbilical connection with the environment long before being aware of it; a connection based on the dialogue of my ancestors with the forest.

I studied engineering a little faint-heartedly, because my thoughts and anxieties lead me elsewhere. My father was pragmatic – he drove me in that direction and I followed. I have no regrets in this regard.

We were part of a Jewish minority in which my father held an important role, not only in Brazil, but also in the world, aiming towards the creation of the State of Israel. Soon after the war, Jewish leaders came to Brazil and met at our house, in Cosme Velho, Rio de Janeiro. I remember very well Golda Meir and Ben Gurion in our living room. During these meetings, films about the concentration camps were screened. I watched everything from a corner, still just a boy. It was the first feeling of institutional violence I experienced which really threatened me.

Later, I had the opportunity to participate in the consolidation of the State of Israel. At that time, we were already aware that investing in knowledge was the most solid pillar for building a nation. And, even today, I keep very strong links with several academic centres in Israel.

My biography has changed several times, and not just from my own decisions, but as an act of nature or, who knows, an act of God. When my father died early in 1957, when I was 30 years old, as the eldest son I inherited the responsibility for managing 25,000 employees. This experience gave me a notion of my own limits. My mother, a Frenchwoman from Belo Horizonte, taught me to be ashamed of effusions and to contain my emotions.

My experience with the movements for the conservation and preservation of nature began with my connection to the WWF (World Wildlife Fund), of which I am one of the first signatories in the Group of 1000. My first social and environmental work was on Ilha Grande, a place with which I feel a strong connection. I helped create ecological zoning concepts resulting in the Tamoios APA (Area of Environmental Protection Tamoios), named after the Tamoios, the first inhabitants of the island.[1]

I visited Antarctica a few times in the 1980s to participate in discussions for the renewal of the Antarctic Treaty – today historical – when claims of sovereignty over the continent were surpassed by common interest, taking as principle the concept of *common heritage of humanity*, originating new forms of internationalization on that continent. I will never forget that white and blue vastness, one of the most striking places I have ever had the opportunity to visit.

I participated closely throughout the formulation of the United Nations Conference on Environment and Development, where I found affinities with the pioneers of environmentalism, such as Maurice Strong, secretary general of the conference, to this day my dear interlocutor in environmental issues.

The conference took place in an unprecedented atmosphere of enthusiasm for what might unfold from that meeting in the city of Rio de Janeiro. I was part of the Brazilian delegation of the Earth Summit, at the invitation of the then Minister of Foreign Affairs Celso Lafer.

Thus was born the concept of "common but differentiated responsibilities," signalling a distinction between developed and developing countries, which today, in a sense, has lost its meaning.

One of the goals of the Earth Summit was the creation of an Earth Charter, a document of ethical principles aiming at a more just and sustainable global society in the twenty-first century. The Earth Charter was an innovative proposal, but it was not accepted by the governments, who eventually adopted another document, the Rio Declaration on Environment and Development.

The 1992 meeting set the participatory nature of civil society on major global issues, NGOs, and social movements being recognized as legitimate.

A HOUSE IN SÃO CONRADO, RIO DE JANEIRO

In this aura of enthusiasm, in that same year the Brazilian Foundation for Sustainable Development, FBDS, was born. Soon after the conference, a new possibility appeared from the idea of sustainable development, words FDBS included in its name. In the early 1990s, the cyclic dynamics in the concept of sustainable development was not as ingrained in our minds as it is today. We decided to make the Foundation a centre for research and applied sciences. In addition, we sought to associate with movements which helped form collective awareness in relation to sustainability. We also sought the co-responsibility of the business sector in their actions, helping to understand how companies are part of the problem and, therefore, must also be part of the solution.

Since those early days, we have acted with the intent to prioritize in FBDS the projects that had influence over the climate changes caused by the energy matrix models. In my new professional destiny I found much more than affirmations: a world of questions.

It was with that concern I took over FBDS, without knowing very well the path I should follow, but instead trying to get education from people who knew more than me. Through questioning, we moved forward. I was cut out for a mission, not for a function. I adopted it as a mission to help rebuild the concept of how to live in this planet.

I took my position as president of FBDS with the commitment to stay only for a while, but I was gradually fertilized by the environmental science and I have been a head of the foundation for twenty years. We needed to have an insight into the historical perspective of the various models leading the planet to the point we are now, and go against the corporate *nouveau riche* affectations related to the decadent and not at all sustainable economic models. Fighting against resistance and helping defeat them has become my daily exercise. And, believe it or not, after twenty years I still face resistance. On this journey, we have warned about the urgency of actions regarding global warming, which is quickly increasing, against the destruction of our natural patrimony, and against the unsustainable use of our non-renewable assets.

FBDS became a partner of the UN, of multilateral agencies like the World Bank and the Inter-American Development Bank, the Global Environment Facility/GEF, and many environmental foundations that care about these matters, such as the Rockefeller Foundation and others. We have worked in harmony with all these groups, within the interrelationship between science and praxis toward new ways of doing.

There were many moments where I felt part of an exploratory group looking for weapons as if it were an *avant-garde*, whose enemies were dangerous perpetrators of environmental crime. This model we felt we needed to overcome was the result of perverse economic choices, committed to an unfeasible energy matrix. There has never been an easy solution. Political decisions take time and do not adjust themselves to the timeline of the current crisis.

Gradually and in a broad context, groups of thinkers from different disciplines allied themselves to the tribe of environmentalists.

It was fundamental to have an impact on public opinion. The reports of the IPCC (Intergovernmental Panel on Climate Change), since the early 1990s, have been of the greatest importance. They presented models, gave evidence, and offered us a prospective view of the approaching crisis.

MY GENERATION SAW BRAZIL AND THE WORLD THROUGH A KEYHOLE

Memory has an ethical dimension. Each one of us always looks for the meetings necessary to our own searches. My generation saw Brazil and the world through a keyhole; we lived in a reality where there were small islands of knowledge. It was also a moment of anachronistic ways of governing. We copied models without having the structure or the culture needed to implement them with positive effects for our country's development. Our own geography signalled that there was a lot to be done – there were huge gaps. Brazil was more like a geographic region than a nation. We had a hint of artificiality, but at the same time we were in a relatively comfortable grandstand, with a good view of the global scenario and some idea of what needed to be done.

The enthusiasm led me to approach a group headed by Helio Jaguaribe. My long friendship with Helio was catalysed by our dear friend, the poet Augusto Frederico Schmidt. This occurred between 1949 and 1950. Young people were looking for gurus from within the intelligentsia in Rio de Janeiro at that time, and we elected Schmidt our leader. He was a permanent presence in my life until his death, as well as San Tiago Dantas and a few others.

All of us had an inherent curiosity about the same subjects. Little by little we expanded our reach. Then came the idea of a conversation with the editorial of the newspaper *Jornal do Commercio*, where a page on Sundays was offered to us in the culture section. Thus was born "Página Cinco" ["Page Five"]. But we needed to go even further. We had to publish, and therefore, we published. There were only two editions: "Ensaios" (Jorge Serpa and Helio Jaguaribe) and "Poemata" (Oscar Lorenzo Fernandes, José Paulo Moreira da Fonseca and I). Was I a poet who felt inhibited by the shame to rhyme?

My anxiety for communication increased exponentially. Helio Jaguaribe then built a bridge so that a group of political and philosophical thinkers from São Paulo could join us. The chosen name would represent somewhere in between our two cities, and thus was born the Grupo de Itatiaia. There could be no better place: the need for action in this group was combined with a landscape in which the nature was as fertile and extensive as the ambitions we had for our country. There was a total of five meetings, held with great effort by the small group. But the fact is that we were looking for directions for Brazil to take.

The highlight of this stage was the first recognition of our "dangerousness" via bitter articles in *Tribuna da Imprensa*, denouncing us with accusations that were so irrelevant I cannot even recall them. What I do remember was a reverberant article by Carlos Lacerda about our group of thinkers, calling us communo-fascists – we were anything but that.

My first look outside domestic circumstances was in the early 1950s, a time when travels abroad were rare and, for this very reason, seen as important events. I chose Paris for a graduate degree at Sciences Po.[2] I remember the loving reception offered to me, an almost unjustified welcome by João Guimarães Rosa, who served in the Brazilian embassy. I, a shrieking but eager student in Paris. Among the memories of that period, I recall squids in a Tunisian restaurant and a conversation, almost a discovery, about salvation and anatomy! Rosa's very well fed smile forced me to work violently and for a long time on my project.

Upon my return from Paris, I was invited by Roberto Campos, then economic adviser of the Joint Brazil-United States Economic Development Commission. Roberto was my first boss, but soon became a great friend.

We have faced many battle fronts together: Brazil-U.S. Commission; BNDE; Roberto, Ambassador in Washington, and I in the team of the Alliance for Progress in the White House; Brazilian foreign debt negotiations; our trips to Europe together; our solidarities, metaphysical and political anguishes about the destiny of our country; our common friends San Tiago, Schmidt, Dr. Bulhões, Mário Henrique, Helio Jaguaribe and a whole generation that made this country. We exchanged, in decades of friendship, a lot of correspondence about the destiny of Brazil.

The joint commission lasted two years (1953-1955), and the development project became a central theme for government policy formulation, with the building of knowledge to enable infrastructure creation for industrial development. The main goal was to finance modernization programmes for infrastructure sectors such as energy and transport. Thus was born the BNDE (Brazilian Development Bank, later BNDES).

It was in my house, in 1954, that Schmidt introduced Juscelino Kubitschek to the intellectuals who had founded the Grupo de Itatiaia, which would later become the ISEB, Higher Institute of Brazilian Studies.

One of the tasks assigned to me, at the invitation of President Kubitschek and at the request of Roberto Campos, was the participation in the Grupo do Nordeste and, later, in the creation of Sudene in 1959, an autarchy subordinated to the presidency. Celso Furtado was its executive secretary.

When President Kennedy was elected, I was invited by the former U.S. Ambassador to Brazil Adolf Berle, then U.S. State Department Assistant Secretary for Latin America, to join the Alliance for Progress. My frequent trips to Washington and New York, besides the contact with the group that focused on that project, opened the door for me to a new environment of *intelligentsia* revolving around the so-called Camelot, which was nothing more than Kennedy's court. At the height of the Cold War, I remember very well when John Kennedy intercepted Soviet ships during the missile crisis. I was having dinner at the Cosmos Club in Washington, with Alexandre Kafka, master of a generation of Brazilian economists, and the President of the Central Bank of Australia, Herbert Coombs. We sat in front of the television to hear Kennedy's speech when he declared that Soviet ships bringing weapons to Cuba would have to return or be sunk. I only remember Dr. Coombs' words: "Don't you think that it is time for us to go home?"

My generation grew up in the shadow of impending nuclear war. The Cold War also brought another main player: the competition for productivity in its models. Which was the most efficient model? What did it mean to be more productive? Was it the one that best absorbed the available workforce, the one that was capable to produce a surplus of national income, which, in turn, would be reinvested in economic development? Anyway, the environmental issue was not on the agenda of any ideology.

Some think the embryo of the environmental issue appeared in the 1960s. But in reality, there was an awareness emphatically focused on the achievements of the model elaborated in Bretton Woods in 1944.

During the Golden Age (1949-1973) – a term coined by the historian Eric Hobsbawm – environmental destruction drew little or no attention.

Sustainability for me is the convergence of many vectors because it is based on dynamic and multidisciplinary sciences evolving in the confrontation between observing the past, knowing the present, and seeking the necessary balance to build a better future.

I am often questioned in interviews about what it means to be an environmentalist. I usually answer that if you want to make an environmentalist glad, just announce a new increase in the oil price in the international market.

All parameters within which we learned how to live have shaped our social, economic, and political relations, from which we assume two fundamental aspects: the sustainability of these models and the ability to adapt to be sustainable in the future. Is that possible? This is what we will examine throughout this book.

In the winter of 2010, I invited Cristina e Leonardo – my friends and collaborators in the preparation of this book – to spend a period of immersion in my reservation in the Pantanal in order to plan and organize our goals for the content of this book. The stimulating environment of the Pantanal would help us focus on my goals. In that atmosphere of serenity, we met in my office, isolated from everybody else. Our tasks were only interrupted when my wife Lea alerted to the arrival of a distinguished visitor. A Jabiru approached our house now and then. We stood in respectful silence watching it coming, in its haughty walk to the veranda. It was as if the Jabiru knew that it was the owner of that territory, a wading bird, with a calm way of walking but a heavy flight. In that winter of 2010, the cold was unusual; temperatures reached five degrees. For fifty years I have been often spending time in the Pantanal, but I had never experienced such rigorous cold. Perhaps nature was trying to tell us something.

Since my youth I have asked the sages of the Bible, the poets, Nietzsche, Schopenhauer, Kant: what I am doing here? Now I will return to one of the great ones. I started this presentation with Fernando Pessoa in one of his pseudonyms, Alberto Caeiro: "Beyond the bend in the road". Now I will let the poem speak: "Beyond the bend in the road/ There may be a well, and there may be a castle,/ And there may be just more road./ I don't know and don't ask."[3]

The "I don't know" has always been for me the first step in the direction of so many questions. A new science combines specific knowledge, so juxtaposed to produce a tool to view the world differently. What will be our legacy? What kind of individuals do we want to be? These are the two great issues of our time. The contemporary debate is about how to change practices and values.

I shall close with a poem by T. S. Eliot, another old master who has been with me for decades, an essential writer for understanding the dilemmas of man.

BURNT NORTON

Time present and time past
Are both perhaps present in time future
And time future contained in time past.
If all time is eternally present
All time is unredeemable.
What might have been is an abstraction
Remaining a perpetual possibility
Only in a world of speculation.
What might have been and what has been
Point to one end, which is always present.[4]

Dialogue in two periods

THE ACCELERATION OF HISTORY

My emotional and intellectual background has been developed thanks to my "old fellows," as I like to refer to those who nurtured my life. Here I shall return to Schmidt, a prestigious literary figure. Each of us who enjoyed his company saw him as a reflection of our best qualities. I was a boy, he was in his 30s.

In Petrópolis, at my parent's house, Schmidt forced me to listen to him recite Camões and his own poems. I remember business meetings in companies where we were co-directors. Unnecessarily long and boring meetings in which he handed me vicious notes referring to the "serious" words and "deep" concerns in the expressions of other directors.

And why do I remember Schmidt now? It was from him that I received a warm letter of recommendation to visit Daniel Halévy, while I was studying at Sciences Po in Paris. Halévy had a great influence over me for his understanding of youth. It was the moment of existentialism and revalidation of Heidegger. Halévy was an old Jew with a long beard who liked to meet with young students and promote lively debates about philosophical questions. He invited us to his home on Fridays. We sat on the floor to listen to him.

Writing a book always reminds us of other books and, in this way, I met again with Halévy in *Essai sur l'accélération de l'histoire* (*Essay on the Accelleration of History*), in which he reproduces a dialogue from October 5, 1946, with the title *L'Histoire va-t-elle plus vite?* (*Is History Accelerating?*). Around Halévy there was also the figure of Raymond Aron.

As we get older, we find a new meaning to the idea of freedom and are able to take more risks. It is through this door I decided to approach conversations between Halévy and Aron. They in 1946, I in 2011, as if, again, I saw myself seated on the floor at Halévy's. I needed to tell them how our world is going.

Halévy's and Aron's words below are a literal transcription of what was said at that time. Below, an imaginary dialogue in two different periods (1946 and 2011).

– Master, the present screams with increasing urgency, and you were right: the acceleration of historical time occurs at intense pace. This is the world today: we are almost seven billion people on Earth, mostly concentrated in urban areas. There is an increasing demand for natural resources. Our models, believe it or not, are still the same created at Bretton Woods in 1944, which you witnessed. Michelet's thoughts, on which you dedicated many studies, already pointed out in the nineteenth century the fact that the pace of historical time had changed. Michelet himself experienced one of the most troubled moments in the History of France, having seen and taken part in three revolutions, in 1820, 1830, and 1848, and his life and work were marked by this acceleration.

Halévy – Indeed, Michelet's words *"l'allure du temps,"* which have the charm and character of his writings, induced me to reflect on the theme we are discussing today. The movement of historical time is not uniform. One need only think for a moment. Time does not have the same value in Naples or in New York. Each history has its own volume and pace. Today, this idea touches us even more deeply, because we are living through a technological revolution, which makes evident the acceleration of time.

– Master, at this moment I am sixty-five years ahead of you in time, and I tell you: I am watching an unprecedented acceleration of history. We go forward with technological tools to the point where we can be anywhere on the planet in real time. Scientists are now able to measure the damages caused to the balance of the Earth. But there are also things that have not changed: oil and coal remain the biggest protagonists of the productive gears of man in the twenty-first century.

Halévy – I confess, quite frankly, I always remembered with sympathy what Renan said about the Exposition Universelle of 1855 in Paris:[1]

"So many things I can do without!" His declaration seems very appropriate to me. It would be possible to live without many things. However, one should not exaggerate, and I do recognize everything we owe to technological progress, which can be of great help if it remains like that. Technology has been and will be useful in the future; it is very positive as long as it is used to serve us. But I do not enjoy it when it becomes a troublesome servant.

– How up-to-date your words sound! Until the mid-twentieth century, nature still threatened man. With technology, man began to threaten nature. What is clearly happening is that we have to accept various impacts. Diasporas will not only be ethnic and political – like those your generation witnessed. They will be related to the climate. The world spends about 1.5 trillion dollars in armaments and military forces. They call it defence spending. And the world today is not even safer.

Aron – Nowadays there is not only a technological transformation; there is a huge transformation of the social structure. It touches the whole of humanity which, for the first time, has a common history. Inside each society, not only minorities are affected, but the whole.

– Our problems are increasingly problems related to all mankind, and they go beyond social structures, extending to geopolitical relations. We live in what would be unimaginable for both of you: at the same time we cause it, we are victims of climate change, an expression that did not even exist in the writings and concerns of your generation. Today, the earth is telling us with very evident signs that we are facing a huge problem.

Aron – It is possible to live without some things when we are on the privileged side, but, at the same time, there are millions who live in extreme poverty. When we forget what technological progress brought us, we forget that in all places where it is absent, that is, in the multitudes of Asia and Africa, awful poverty reigns. I would live without many things, but I wish there were not so many millions of people without the essential, which only technological progress can offer.

– Today, China takes possession of technological advances very quickly. For thirty years, the Chinese have been going through a great leap forward. But many African countries still live through the consequences of poverty. And what we see is that in many parts of the planet the poor are the first to suffer the effects of extreme weather events, like floods and draughts. Dear masters: it is urgent to think of a new political and economic architecture to guarantee human life. And I will tell you a perhaps surprising fact: if we do not follow different paths, future generations will be under threat.

1

Limits of the Earth

Why are human activities changing the biosphere so profoundly? We are in need of understanding interactions between geophysical and socioeconomic systems—that is the challenge. We must also consider the political and cultural motivations which have led to this reality. A new dimension of knowledge and consciousness is opening up to us, one that could perhaps be the key to our future.

It is thus necessary to better understand how our little home works in order to reorganize it.

We should also remember. I relentlessly maintain a file of all the correspondence, articles and lectures I have given since the 1960s, very well kept thanks to the diligence of my secretary for more than forty years, Carmen Prins. Much of this collection has served as raw material for this book.

For example, at the opening of the 1998 World Bank Annual Meeting I called attention to the fact that

the atmosphere and life are interdependent and have co-evolved. The atmosphere of Venus, Mars and all the other planets of our solar system preclude the existence of life. A thin blanket only 12km thick surrounds our planet. Its composition is indispensable, having enabled the processes of evolution and the survival of life on Earth. The stability of our atmosphere is maintained by this long, complex, and extensive ecological interaction.

The end of many species is now an unremarkable fact, and the present increases to the extinction rate are the result of a process caused by the improper and careless occupancy of spaces that do not belong to homo sapiens. *By persisting in his role as the disrupter of ecological interactions, man becomes his own predator, destroying his source of survival.*[1]

The ancient Greeks developed the concept of *hubris*, the downfall that occurs when man surpasses his own measure; it is likened to an act of arrogance. To the ancients, politics, ethics, and the use of natural resources were questions about balancing between extremes, of restraint before the immensity of a planet they had only just begun to understand.

There are three types of environmentalists: the dreamer, the utopian, and the profiteer. The first builds castles in the air; the second lives in those castles... the third charges rent from both. But there is a fourth type of environmentalist, the one who builds with the raw material science provides him, aiming towards an environmentally viable future for generations to come.[2]

A RECENT HISTORY

Due to the complexity of ecosystems, ecology needs to be supported by several sciences, such as climatology, physics, biology, chemistry, and mathematics, as well as anthropology, sociology and psychology. An intimate dialogue among all these parties strengthens the whole. Ecology extrapolated this scientific field and connoted political and ethical positions about the relationship between man and his home. Therefore, the discipline of ecology imposes several constraints.

Environmental degradation only reached the international negotiation table in the 1970s, during the Stockholm Convention, which presented something new for that time. The Stockholm Convention concentrated on the contamination of chemical waste and the pollution caused by industrialization—problems of the First World.

From the Convention of 1972, international financial organisations and institutions tentatively began to associate environmental demands with policies arguing for the release of capital for development agendas.

Development meant economic growth at any cost. Brazil even made a statement after Stockholm as follows: "We want developed countries to send us their pollution, that is, bring their industries here."

Stockholm stood out less from its practical results and more from its heated debates, strongly influenced by a report from 1972 commissioned by the Club of Rome entitled "The Limits to Growth." This report, organized by a group of researchers from MIT led by Dennis Meadows, showed that the planet would not be able to support demographic and industrial acceleration due to pressures on natural resources. It would be necessary to freeze population growth and the stock of capital in order to achieve economic stability and respect the limits of natural resources. The Meadows Report was one of the first serious attempts to predict the future through mathematical modelling. The keyword of that work was *balance*.

The study quantified the reserves of many natural resources, such as copper, oil and cobalt, and calculated how long it would take for those reserves to be depleted. Subsequently, those calculations were strongly criticized for not having incorporated technological advances which would have permitted exploration under different conditions so, therefore, the results were seen as invalid. Today, the report's basic theses have been retaken within the context of new technologies and further advanced knowledge.

The position of the Meadows Report was linked to the old Malthusian arguments stating that population growth occurred within a geometrical progression, while food production grew arithmetically. Therefore, without birth control, hunger or war would be the inevitable outcomes. The book *The Limits to Growth* sold 30 million copies, becoming a best-seller. Some passages sound awfully up-to-date:

"We have mentioned many difficult trade-offs in this chapter in the production of food, in the consumption of resources, and in the generation and clean-up of pollution. By now it should be clear that all of these trade-offs arise from one simple fact—the earth is finite. The closer any human activity comes to the limit of the earth's ability to support that activity, the more apparent and irresolvable the trade-offs become. [...] In general, modern society has not learned to recognize and deal with these trade-offs. The apparent goal of the present world system is to produce more people with more (food, material goods, clean air and water) for each person. In this chapter we have noted that if society continues to strive for that goal, it will eventually reach one of many earthly limitations [...] We are unanimously convinced that rapid, radical redressment of the present unbalanced and dangerously deteriorating world situation is the primary task facing humanity. [...] This supreme effort is a challenge for our generation. It cannot be passed on to the next. [...] We have no doubt that if mankind is to embark on a new course, concerted international measures and joint long-term planning will be necessary on a scale and scope without precedent. [...] We affirm finally that any deliberate attempt to reach a rational and enduring state of equilibrium by planned measures, rather than by chance or catastrophe, must ultimately be founded on a basic change of values and goals at individual, national, and world levels."[3]

Twenty years have passed and we organized the Earth Summit, calling for the same urgency. In 2012, year of Rio+20, the Meadows Report will be 40 years old. The limits of sustainability for the Earth, only projections in the report, are now an increasingly glaring reality.

What was the Club of Rome, the sponsor of the report? The Meadows Report was brought about through the motivation of Aurelio Peccei, an Italian in charge of Fiat and also one of the founders of the Club. Scientists, economists, entrepreneurs, educators, and bankers gathered at the Accademia dei Lincei, in Rome, driven by the enthusiasm of Peccei, aiming to point out ways for solving problems like population growth, poverty, urban sprawl, and pollution. The Club of Rome defined itself as an informal school and, in its early years, Helio Jaguaribe was one of our representatives, my brother-in-arms who travelled with me over so many roads in Brazil for more than half a century. Helio is, until today, an honorary member of the Club, as well as Fernando Henrique Cardoso.

At that time, the Meadows Report was accused of catastrophism. A more optimistic view emerged in the 1980s, with the World Commission on Environment and Development, established by the UN and chaired by the Norwegian Gro Brundtland. In the report "Our Common Future," we see for the first time a definition of sustainable development: "development that meets the needs of the present without compromising the ability of future generations to meet their own needs."[4]

Sustainable development is a vital yet contradictory concept. Etymologically, *desenvolver* [in Portuguese] means to undo what is *envolvido*; in Spanish, *desarrollar* that which is *arrollado*; in French or English, *development/développement*, that is, *disenvelop*, to permit the release or appearance of something that was hindered or hidden. The notion of development involves dynamics and therefore movement. The notion of sustainability, in turn, implies a static situation which presupposes permanence. Economic development is aimed at improving living conditions, but implies an impact on nature. Sustainability is based on the idea of the balance and preservation of the environment. This equation challenges us every day. We still do not know how to measure precisely all vectors of the concept of sustainability.

There is widespread misappropriation of the expression "sustainable development," as it started to be used in projects which were not at all sustainable. The ethical primacy of the term was lost gradually over time. We need to recover it.

THE EARTH SUMMIT IN RIO DE JANEIRO

In 1992 we still did not know exactly what the concept of sustainable development meant, but we knew it would lead with an ethical perspective. For us environmentalists, the Earth Summit [also known as Rio-92] inaugurated a new phase. We were impatient for actions to save the planet.

World leaders admitted there were major problems. It was a great step, an unprecedented meeting in the history of the UN, both for its size and for the scope of its accomplishments. While heads of state gathered at the Riocentro convention centre, NGOs set up their campsites in the park of Aterro do Flamengo. The Earth Summit had a huge impact worldwide, with the presence of over ten thousand journalists.

In December 1989, preparation for the summit began, involving planning, education, and negotiations among UN member states, as well as the adoption of principles that later would be part of Agenda 21, one of the documents approved in 1992. The agenda is a comprehensive guide of actions aimed to achieve sustainable development in the 21st century – hence the title, Agenda 21 – in which we have now been living for over ten years. The agenda was never implemented, but entered the history of environmentalism for its symbolic and educational characteristics. In its 40 chapters dealing with several areas of human and environmental development, I would like to highlight:

"Scientific knowledge should be applied to articulate and support the goals of sustainable development, through scientific assessments of current conditions and future prospects for the Earth system. [...] In the face of threats of irreversible environmental damage, lack of full scientific understanding should not be an excuse for postponing actions which are justified in their own right. The precautionary approach could provide a basis for policies relating to complex systems that are not yet fully understood and whose consequences of disturbances cannot yet be predicted."[5]

The conference took place in the first two weeks of June. Days before, I was given the task to gather a group of businessmen at Marina Palace Hotel for a presentation on sustainability and on what the Earth Summit would be:

the world we have today is a world of infinite possibilities for the creation of something better, or for an incredible danger of catastrophe. We must realize that the priority given to these issues, in practical terms, is not being accompanied by decisions involving urgency, and that we all consider it necessary for this matter to be resolved.[6]

I saw skeptical faces in the audience. The businessmen were resistant to environmental issues throughout the Earth Summit. In the 1990s, we were little more than a few hundred environmentalists; we formed a small global community and we all knew each other. Some disappeared along the way. The massive mobilization that exists today around environmental issues started with the ideas of our team of old combatants.

The Secretary-General of the Earth Summit, Maurice Strong, one of those combatants, used to tell us that the conference would not be a silent one, either by success or failure. As Maurice recalled:

"I watched them filing into the room – heads of state, presidents and kings and prime ministers, tyrants and democrats and dictators and builders of consensus, men (and a few women) who represented all the teeming billions on the planet. [...] Slowly they took their seats at the huge oval table we had constructed specially for them. Oval because ... well, bringing 116 leaders together in a room poses delicate problems of protocol, as you can imagine. That table was one of the logistical challenges for the conference—you couldn't have anyone appearing to take precedence."[7]

In 1992, for the first time, the role of civil society was recognized. Until then, only governments represented society. This mindset was a consequence of Cold War geopolitics and centralized state decisions still prevailed. A new voice appeared at that time, called NGOs (Non-Governmental Organizations). There was an aura of enthusiasm, but the economic model was not discussed at any point. Just as, indeed, it was not discussed in subsequent conferences.

The Earth Summit left us two important comprehensive multilateral treaties, which today still form the institutional framework used in environmental negotiations at the UN: the Climate Convention and the Convention on Biodiversity. Thereafter, each convention followed its own path, with varying degrees of success.

The Climate Convention is the legal benchmark in which everything about the climate is discussed. It is a treaty between countries which, although at the time of signing had not established limits on the emission of greenhouse gases (GHG), had historical importance and remains as the framework of ongoing negotiations. It is the official recognition that anthropogenic emissions of GHG had been affecting the Earth's climate, and that rich countries should reduce their emissions to "non-dangerous" levels. What does "non-dangerous" mean? This level should be reached within a time frame sufficient to allow ecosystems to adapt to climate change, ensuring food production, providing sustainable economic development. Common responsibilities were established between rich and poor, the definition of which were given to the countries who became part of first Annex of the UNFCCC. These nations became the leaders of the process, namely the fight against climate change.

It was up to the Kyoto Protocol, five years later, to set up goals and make the distinction between industrialized countries (Annex I) and non-industrialized countries (Annex II).

We saw the definition and results of being "developed" were under a limited umbrella, as they did not include the social factor, let alone the environmental one. It was necessary to reach the highest philosophical level for the equation to complete itself.

THERE IS NO SINGLE *EUREKA* MOMENT

The Englishman John Elkington brought the term *triple bottom line* into the history of sustainability, integrating the social to the environmental and the economic. In that tripod all legs should interact and have the same strength. Otherwise, development would not be sustainable.

The strength of Elkington's formulation comes from the fact that it presented a new language, so civil society and companies were able to include environmental values in their bottom line. Until the 1990s, many groups and environmental movements used a politicized language, sometimes incorporating old Marxist concepts typical of the Cold War era. This kind of discourse would hardly raise public awareness. A new language and a new vocabulary were necessary in order to help people understand and get involved with environmental issues.

When asked how he had coined this term, Elkington answered that there was no single *eureka* moment: "We felt that the social and economic dimensions of the agenda-which had already been flagged in 1987's Brundtland Report (UNWCED, 1987)–would have to be addressed in a more integrated way if real environmental progress was to be made. [...] Like Paul McCartney waking up with *Yesterday* playing in his brain and initially believing that he was humming someone else's tune, when the three words [*triple bottom line*] finally came to me I was totally convinced that someone must have used them before."[8]

The traditional *bottom line* of a company is profit. If there is profit, the company is doing its job satisfactorily. The concept suggested by Elkington added two elements to this assessment basis, giving equal weight to economic, environmental, and social aspects. Elkington, later on, would translate his concept in the expression "3P": *people, planet, profit.*

Public awareness grows as people see the degradation of the environment they inhabit.

Thinking about sustainability also implies the management of decisions that take into account short, medium, long and very long terms. The purpose is the maintenance of life, especially human life, on the planet. The concept of sustainable development is increasingly expanding, where concepts from economy are translated into ecology and vice-versa.

Economic deficits are those we borrow from each other. Ecological deficits are those of future generations and we do not know how to return them. When we were few, the capital produced by men was scarce. Now, the natural capital provided by ecosystems is scarce, a limiting factor to the quality of human life on Earth.[9]

The concept of natural capital is complex and still under debate as we acquire more knowledge about ecosystems. It is not, for example, a mere question of counting the number of trees in a forest and calculating the value of their timber. Natural capital means natural wealth from the physical processes that support the ecosystems. In this equation, we include broad aspects of ecosystems, like biological stocks and the regulation of water supplies, as well as the absorption of greenhouse gases.

Natural capital presents itself in two forms: the critical and the renewable (or replaceable). Critical natural capital is essential to the maintenance of life and to the integrity of ecosystems, and its depletion is disastrous. In turn, renewable forms of natural capital are those which can be recovered or replaced, such as a portion of decertified land undergoing reclamation, or the constant flow of solar energy we can use to generate electricity. Human activities affect the natural capital in different ways—they can maintain it, decrease it, or increase it. Having knowledge about the effects of economic activity on the real-world level of local ecosystems is one of the preconditions for achieving the environmental *bottom line*.

What about social capital? What does it mean? This concept is based on the notion where social networks are assets with value. A society in which people have a sense of belonging to the community generates more value than in one which does not. Trust and tolerance are some of the elements that make common coexistence possible. Societies with a good stock of social capital enable and benefit from low crime rates, better health, education, and higher rates of economic growth. According to Robert Putnam, a political scientist from Harvard, whose theories have gained prominence over the last decade,[10] social capital refers to the connections between individuals, to different forms of association and civic participation. These interactions permit individual talents and virtues to spread and expand. In this sense, a society with many talented yet isolated individuals is poorer than another society whose structure is strengthened by close social relations. A society whose members are connected politically and emotionally is able to promote a better quality of life than a society where people are isolated within their individualities. Hence the importance of the social standing in the *triple bottom line,* which promotes new forms of participation and association for a better quality of life.

Sustainable development can be understood as a balanced integration between society, economy, and nature, enabling the permanence of ecosystems and the generation of wealth for all. It is not easy, but we are preparing the groundwork for this to occur.

RIO+5, AN INTERLUDE

The year 1997 was marked by the 3rd Conference of the Parties in Japan. Here the Kyoto Protocol was signed, the first international treaty in which the signatory developed nations committed to reduce their emission of greenhouse gases. At the end of the conference in December, the protocol was officially opened for the signatures of the countries. The United States, the biggest emitter of GHGs at that time, signed it, but never ratified it. Kyoto was a test for multilateral negotiations, consolidating the *polluter pays* principle, and created a worldwide carbon market. Today, almost fifteen years later, we know well that its main goal—the stabilization of greenhouse gas levels in the atmosphere—was never reached. Quite the opposite: emissions are increasing at a worrying pace.

All COPs have basically two guiding lights. Namely: the Climate Convention, of 1992, and the Kyoto Protocol, of 1997. The convention is hierarchically above the protocol. Today, we question if this legal framework is sufficient or even appropriate to solve the climate issue. The world has changed, China and India have grown, we are living in a different reality. But what did not change was that the United States has neither reduced its consumption and nor has it shown any interest in leading the process of emission reduction.

Earlier, in March of that same year, the Rio+5 Conference took place, once again in Rio de Janeiro. The meeting could never have happened without the leadership of Maurice Strong, who chaired the Organizing Committee with me. It was the first initiative to evaluate the results of the Earth Summit. The conference was attended by social movements,

NGOs, indigenous peoples and a political celebrity: Mikhail Gorbachev, the man who ended the Cold War without firing a single gunshot.

In retrospect, 1997 was an eventful year. My inaugural address at Rio+5 brings me good memories of the hope that lingered in the air.

The title of our project is Sustainable Development. Its mechanism is Agenda 21. Since 1992, the emphasis on natural resources has expanded due to the understanding that the environment and well-being are interconnected. The prerequisite for sustainability and social needs has made us have a critical view of our social, political and economic models.

Will the dream of freedom and equality be accomplished without solidarity? Will it be possible to achieve solidarity without a change of lifestyle and without the renewal of ethical values? Definitely not!

The most important obstacle is perhaps the success of the current economic model that tied permanently the creativity of a market economy to technological advancement. Paradoxically, this model produced relative inertia in relation to the inclusion of environmental and social vectors on its agenda. Another by-product of the success of the modern economy is a global doctrine that probably does not meet the needs of those who do not enjoy the benefits of economic and technological progress. We must stand together to ensure the dignity of that huge portion of humanity that still does not have access to this benefit. This is the essence of sustainability. The political model is lagging behind, still restricted to the limits of national states or even to tribal and ethnic groups. The world, however, is global, and new forms of government will emerge from current conflicts.[11]

Often there was a chaotic atmosphere, with more than 500 people holding different postulations on different topics. I faced resistance from a few NGOs. It is Maurice who well recalls that moment:

"Many of the leading representatives of the NGOs, always prickly about their independence, bridled at the prospect of participating in a process dominated by Klabin, who they didn´t think was sufficiently respectful of their crucial role. There were many tense and anxious moments"[12]

The discussions were indeed tense, but very positive as well, since they revolved around a contradiction between environmental degradation as a threat to society's basis for survival and the actions of governments and companies. The social element had not been emphasized in 1992, and both Maurice and I wanted to bring it forth to the agenda of Rio+5. We were convinced that it was important to move systemically.

During Rio+5, the Independent Earth Charter Commission, created one year earlier by Strong and Gorbachev, convened its first meeting and at the end of the conference a reference text was released as a "document in progress," taking aim towards a more sustainable global society. In the Charter text there is a section called "Earth, Our Home":

"Humanity is part of a vast evolving universe. Earth, our home, is alive with a unique community of life. The forces of nature make existence a demanding and uncertain adventure, but Earth has provided the conditions essential to life's evolution. [...] The global environment with its finite resources is a common concern of all peoples. The protection of Earth's vitality, diversity, and beauty is a sacred trust."[13]

Here I will make a parenthesis: Ruth Cardoso attended a seminar at Rio+5. I remember at her side was seated the princess of Jordan—today Queen Rania. On March 13, at the opening of the conference, I woke up to the news that a Jordanian soldier from Eilat, between the border of Jordan and Israel, had shot at a school bus filled with Israeli girls. And there I was on that day, sitting beside the princess, born in Palestine. We ended up commenting on the conflict in the Middle East and it was a positive dialogue, as we discovered that we shared the same opinion about peace. At that moment, this event went beyond the theme of the conference.

DEVELOPMENT CAN BE CLEAN

After meetings and discussions with several sectors of civil society, representatives of minorities and indigenous peoples, it was time to work on a systemic project that contained a seed of change.

That year it was important to give technical support to the formulation of a theoretical framework which would be consolidated, after much negotiation, into the Clean Development Mechanism (CDM). FBDS was one of the institutions that worked on this project, which was eventually taken to Tokyo by the Brazilian delegation in December 1997.

We environmentalists had an increasingly clear idea that it was important the goals for industrialized countries were anchored on the principle of historical responsibility, i.e. countries which have introduced carbon into the atmosphere since the Industrial Revolution held greater responsibility in relation to increased global temperature.

Through the principle of "common but differentiated responsibilities," developed countries came to have binding goals for the reduction of emissions, i.e. they were obliged to reduce them to 1990 levels. Each country internally decides about the allocation of goals between their productive sectors. What happens is, for many companies from developed nations, the cost of the emissions reduction is higher than the same reduction in developing countries. As the goal is to contain the overall level of emissions, it was possible to create flexible mechanisms so that emissions in a rich country could be compensated by the emissions in a poor country.

The CDM, one of those flexible mechanisms, allows a company to invest in projects for the reduction of emissions in developing countries. What the company from a developed country does not reduce from its quota of emissions can be compensated with reductions in a developing country. The CDM established the technical parameters for such compensation. These projects have to be voluntarily adopted by developing countries and are conditioned toward a real and measurable reduction of emissions in the receiving country.

The CDM has become a reality, yet global emissions of greenhouse gases are increasing because of flaws in the economic and political models, as we shall see in the next chapters. Obviously it is not possible to overcome the problem in one fell swoop.

Following on from the flexible mechanisms destined to support the reductions of GHG, I helped in the structuring of the first carbon stock exchange in the world, the CCX (Chicago Climate Exchange).

The CCX operates as a *locus* for the commercialization of carbon credits, helping to lower the costs of companies in this kind of transaction. A company which has reduced its emissions beyond the targets can sell this surplus reduction to another company that has failed to reach their reduction targets. In this manner, the average global cost of the reduction of emissions could be reduced.

I participated in the creation of the CCX, together with Richard Sandor, CEO of the exchange. I was part of the Advisory Board of CCX since before the beginning of its operations, in 2003. The carbon credit market is a kind of *cap-and-trade* mechanism which could be significantly important in reducing greenhouse gas emissions.

We were the first institution to submit projects outside the United States for the CCX. The Chicago Climate Exchange ended up empty after the economic crisis of 2008/2009, and its control was bought by the stock exchange of Atlanta. However, the business model that structured the CCX system was replicated in many parts of the world and still helps in breaking down the costs of reducing greenhouse gases. "Much ado about nothing," as Shakespeare well said? No, because the road to sustainability is built this way, with experience, mistakes, and successes.

In March 2011, driven by a desire to talk about our dreams back in 1992, I called Maurice Strong, who now lives in China.

CONVERSATION WITH MAURICE STRONG

– Dear Maurice, I miss you. You have not only the wisdom of the elders, but also the foresight of ancient prophets about the future, both from my people and from the Chinese.

Maurice, we have certainly worked together for more than twenty years. We did not change, but the world has changed. In 1992, we thought the same way we think today. Next year, 2012, is the 20th anniversary of the Rio Earth Summit. Are you satisfied with the progress that has been made since then?

Strong – I am not at all satisfied – indeed, I am very disappointed. Although there are many individual examples of progress that have been made, overall, the Earth's environment has worsened since 1992 and efforts to reverse this trend have reached a low ebb.

– The Principles adopted at the Earth Summit, the conventions agreed and the approval of Agenda 21 as the continuing program of action and the supporting agenda agreed by the non-governmental community constituted a promising pathway to a sustainable future. Why, then, have the hopes and expectations of 1992 not materialized?

Strong – The Earth Summit was one of those rare international conferences in which governments took commitments beyond what they had intended to do when they came to Rio. The reason is the effective and concerted way in which the representatives of civil society came together with broad public support to insist that their governments move on the issue that would determine the future of human society. You know this as you were my principal and most effective partner and it is a privilege as we approach the 20th anniversary to reflect with you of what we did and failed to do in 1992.

– Surely we must use 2012 as a unique opportunity to provide renewed impetus to the transition to a sustainable future for which there was such universal agreement in 1992 – the "change of course" called for by world business leaders in their report to the conference. We have not "changed course" although there have been many changes in the world since then. How do you think this will affect the prospects for a renewed commitment to sustainability in 2012?

Strong – The political situation has changed radically as the centre of gravity of the world economy has shifted to Asia, notably China. This has been accompanied both by greater pressures on the environment and a greater commitment, particularly by China, to environmental protection.

– We had some great moments in Copenhagen. The COPs are ultimately a *déjà vu*. The themes are the same, the excuses of those who do not understand the urgency are the same, and the problem... becomes bigger each time. You know very well that, without the participation of China and United States, there is no possible solution for that which worries us. Maurice, how do you see China nowadays in relation to everything we believe in? What are the prospects that China will accept a new international agreement on climate change?

Strong – The remarkable growth of its economy has made China the largest single source of greenhouse gas emissions, while still on a *per capita* basis, being much less than those of the United States and other industrialized countries. However, China cannot be expected to accept a new agreement on climate change that allows those responsible for the emissions that have produced this crisis to shift a disproportionate share of the burden of reducing emissions to China and developing countries. It nevertheless is implementing measures domestically that go beyond what most other nations are doing as, for example, by imposing limits on automobile emissions greater than those of the United States.

– What, then, do you think that Rio + 20 can achieve on the key issue of climate change?

Strong – Not agreement, but progress towards agreement, and commitment to a continuing and accelerated negotiating process. Efforts to negotiate a new agreement have floundered on the deep differences between the industrialized countries, largely responsible for the accumulated emissions that produced the current crisis. Developing countries with the strong support of China and India understandably insist that those countries responsible for the crisis accept the primary responsibility for stabilizing emissions and the costs of doing this. Sadly, the United States has abandoned the leadership role it demonstrated in Rio and then in Kyoto. This tragic impasse will figure prominently at Rio+20 which is unlikely to resolve the north-south differences, but must make significant progress towards this. It, indeed, is imperative if we are to respond in time to the greatest security threat the human community has ever faced.

– I share your concerns, but remain somewhat more optimistic. Don't you think that citizen power could again be mobilized to drive the political process and ensure that governments at Rio 2012 go well beyond the positions they are now taking as they did in 1992?

Strong – Indeed I think it is the only way we can even hope that 2012 will provide the real impetus to agreement and action that is so urgently needed. No one knows more what is needed or is more capable of leading this process than you, Israel. At the same time, we must realize that the commitments by governments, however necessary, are not sufficient. What is important is the implementation of their commitments. From Stockholm in 1972 and since then and in the many conventions, declarations and agreements, particularly those of Rio in 1992, governments have taken commitments, yet failed to implement them. If they had done so we would not be facing this crisis. So it will not be enough for citizen action to drive new agreements or reaffirmation of existing agreements in 2012. There must be a continuing mobilization of public support and insistence that governments implement the commitments they have made, even to the point of enforceable penalties for those who fail to do so. I would like to be able to share your optimism, but I join in undertaking to do everything possible in 2012 and beyond to set us on the change of course which is even more urgent now than it was in 1992.

– It is up to us, old combatants, help shape Rio+20. Perhaps this time we will be able to defeat all delaying bureaucracies.

A SUPERPOWER OF BIODIVERSITY

Biodiversity is rich, and the more diverse the manifestations of life, greater are the chances of survival and adaptation. Brazil has the potential and vocation to lead the implementation of the concept of sustainable development.

The Biodiversity Convention of 1992 defined biological diversity as: "the variability among living organisms from all sources including, *inter alia*, terrestrial, marine and other aquatic ecosystems and the ecological complexes of which they are part; this includes diversity within species, between species and of ecosystems."[14]

The Nagoya Protocol in 2010 regulated the conditions of access to genetic resources and benefit-sharing for the country providing those resources. While in the negotiations on climate there is blatant lack of political will, in the negotiations on biodiversity we are already seeing progress. In Nagoya the "IPCC of biodiversity" was established, named the IPBES (Intergovernmental Platform on Biodiversity and Ecosystem Services), which serves as an interface between the scientific community and policy-makers in order to develop policies for this field.

If life is equivalent to wealth, Latin America is a superpower in biodiversity. "Latin America and the Caribbean, a Biodiversity Superpower," a study by the UNDP/UN (United Nations Development Programme),[15] indicates that while the 20th century brought awareness of the dangers of environmental degradation, the 21st century is bringing an understanding of the real importance of food, water and climate control provided by our ecosystems.

Biodiversity, in the rich ecosystems of the region, is a treasure whose value tends to increase during the 21st century. It represents a comparative advantage that may enhance social development. These resources are at risk if we continue the practice of *business-as-usual,* which will deplete the ecosystems' resources. The study shows how Latin America and the Caribbean may be powers in the new economy for their rich biodiversity, natural capturing and sequestration of carbon, for their vegetation, and for the largest reserves of fresh water on the planet.

Latin America was born from the exploration of the unknown, from the ambivalence between the exuberance of nature and the will to explore, besides the construction of new societies in the midst of these abundant natural resources. Brazil has a privileged position in this scenario. What we have is no small thing.

The region includes six of the countries with the greatest biodiversity in the world, as well as the most biodiverse region in the world, the Amazon. These countries cover less than ten percent of the surface of the earth, but contain approximately seventy percent of all species of mammals, birds, reptiles, amphibians, plants and insects in the world. It is an enormous treasure, much still unknown.

At the OECD Workshop in 2001, "Market Creation for Biodiversity Products and Services," I observed that

the basic question of biodiversity can be formulated on how to deal with the goods and services provided by nature, respecting their basic needs for preservation and including local populations in the economic benefits derived from these goods and services. Biodiversity placates direct human needs: genetic material for improvement of crops and livestock and substances for new medicines. Reducing the biological diversity of one ecosystem reduces its productivity and its ability to withstand disturbances and increases its susceptibility to pests and diseases. On the other hand, the use of sustainable biodiversity not only produces economic benefits, but also promotes the preservation of ecosystems.[16]

RESILIENCE: ADAPTING TO A DIFFERENT WORLD

I like the idea of resilience, an interesting concept that has been used in several fields, such as physics, psychology and ecology. Resilience means, in general terms, the ability to resist and evolve in the face of crises and aggression. In *Resilience Thinking*,[17] Brian Walker and David Salt observe the broadest interactions between biodiversity, knowledge, and society.

For Ecology, resilience is the ability of an ecosystem to tolerate disturbances without collapsing. A resilient ecosystem can resist shocks and recover when necessary. For social systems, resilience means the capacity for forward planning, gaining strength while facing adversity, and envisioning solutions.

We are part of the natural world, depending on ecological systems for our survival and, at the same time, having impact over them. Cities especially depend on a delicate relationship with their surroundings, the places where they get their food, water, and energy. Thus, the study of resilience aims to determine the amount of change that a socio-ecological system can withstand while still being able to remain complete.

Every socio-ecological system has some level of resilience. Some can withstand more aggression, others are more fragile. But there is always a limit and, after too many changes, a system can definitely become corrupted. This is what happens when deforestation is excessive, which makes it impossible to regenerate a forest.

The idea of resilience helps us realize that habitats are not untouchable. The more biodiverse an ecosystem, the more resilient it will be against attacks of human interference. We can consider biodiversity a kind of insurance against climate change and deteriorating ecosystems. But there is a risk of going beyond the limit. Aggression is tolerated only to an extent.

Human activities always generate some level of impact. If these changes were maintained to levels at which natural systems could absorb and recycle them, there would be no problem. However, we are exceeding those limits.

Almost a third of the Earth's surface has been modified for urban, industrial or agricultural ends. Agricultural areas already cover more than one-fourth of the Earth's land, while we use for this same activity six times more clean water than what we naturally let flow in rivers.

The number of species on the planet is in decline; more natural capital is used than the Earth is capable of renewing. By interfering in the cycle of the destruction and creation of species, we are destroying much more than can be replenished. This is a risk we took for our civilization, since we are above the Earth's limit of resilience.

The resilience of the planet made it possible to systematically pollute and degrade ecosystems, with no clear awareness of what was going on. But the Earth does not function at the same pace as the models created by man. Changes do not happen in a linear fashion, and not all effects have one single recognizable cause. Hence the fundamental importance of biodiversity: the more biodiverse and complex an ecosystem, the more dense the web of relationships and services the species provide one another.

In *Collapse*,[18] Jared Diamond emphasizes the centrality of ecology in people's survival. "At an accelerating rate, we are destroying natural habitats or else converting them to human-made habitats, such as cities and villages, farmlands and pastures, roads, and golf courses. The natural habitats whose losses have provoked the most discussion are forests, wetlands, coral reefs, and the ocean bottom. [...] Elimination of lots of lousy little species regularly causes big harmful consequences for humans, just as does randomly knocking out many of the lousy little rivets holding together an airplane."

HOW MANY OF US CAN THE EARTH SUPPORT?

Environmental deterioration is directly related to population size. According to the UN/DESA (United Nations Department of Economic and Social Affairs),[19] between the decades of 1950 and 2000, the average population growth per year was 1.76%. It does not seem much, but this number means that in that half a century, the population grew 144%, from 2.5 billion to 6.1 billion in 2000. It is a galloping growth whose pace, fortunately, has been reduced in the 21st century.

The UN estimates that from 2001 to 2050, the rate of growth will decrease to 0.77% per year, on average. We will keep on growing, albeit at a slower pace, but it is estimated that the annual increment will still be around fifty million people per year. In five decades, we will have the equivalent of two more Chinas in the world. If this pace is maintained, we will be nine billion human beings on the planet in 2050.

The scenario outlined in the previous paragraph is considered the most probable by the Population Division of UN/DESA, but there are chances that other projections will become true, depending on various factors such as governmental population planning, income increases, advances in medicine, and the perceptions of the population itself in respect to the importance of family planning. The biggest population increase will happen in the poor least developed countries, which will represent one third of humanity. Nigeria, the Democratic Republic of Congo, Sudan, Bolivia and Pakistan are the countries that best represent this group. Their populations in general have no access to energy. It is a huge contingent who will require answers to their needs. The larger the population, the greater the demand for water, food, and energy.[20]

In the first half of 2011, the *World Population Clock,* of the U.S. Census Bureau[21] informed that we are already 6.9 billion. Everybody needs food, water, shelter, education, work, and income. But reaching these goals using current models of production and distribution is incompatible with the physical limits of the planet.

The legitimate desire for raising the economic and social share for populations still at the margins of modern welfare standards will only be reached if there is a major review of our prevailing ways of life. Energy and consumption patterns need to be changed so the planet will be able to keep supporting us.

ON TRACES AND FOOTPRINTS

Insofar as public awareness of our environmental situation grows, society demands more information and new ways to think about the problems. Only then will new concepts, indicators, and alternative ways of seeing environmental degradation emerge. One idea that has appeared during the last decade of the last century was the "footprint." The image it evokes helps in thinking about the concept: what size are the traces we leave behind when we perform our activities? How can we quantify, objectively and intelligibly, the extent of damage caused to the biosphere?

The ecological footprint is an attempt to answer these questions; it is an instrument for thinking about sustainability and a way to have an indication of anthropic pressure, that is, how much of the planet's natural resources are consumed by man, individually or collectively.

The ecological footprint measures how quickly we consume resources and generate waste, compared to the pace at which nature can absorb the waste and generate new resources.

Since the 1970s, humanity has been exceeding the Earth's regenerative capacity. Today, the planet needs a year and a half to regenerate what was consumed in one year. As the initial reserves of natural capital were relatively high, it was possible for some time to withdraw more than what could have been considered wise.

The idea of the ecological footprint was created in 1990 by Wackernagel and Rees, from the University of British Columbia, and is now used by scientists, government agencies and institutions monitoring the planet's conditions.[22]

The ecological footprint proposes to define the relationship between anthropic pressures (human impact) and biocapacity (the planet's capacity to absorb waste and generate new resources). But no indicator or model is able to capture the full impact of human activities on the ecosystem.

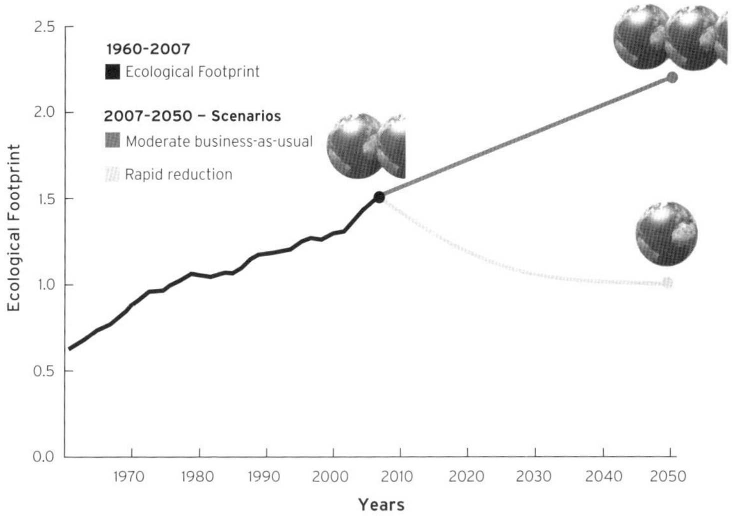

The above graph indicates the world environmental footprint since 1960 up to the projection for 2050. Biocapacity is exceeded when the line representing the footprint goes above 1.0 – which occurred in the mid – 1970s.

Carbon emissions account for more than half of the total footprint (54%) as it is the component that has been growing the fastest. The total carbon emitted in a year is compared with the portion of the planet necessary to absorb emissions.

By combining different factors such as carbon, changes in land use, agriculture, and fishing, the ecological footprint clearly indicates the choices we must make, that is, to consider the trade-offs between basic natural resources. If we want to absorb more atmospheric carbon, it will be necessary to reduce the area taken by agriculture and urbanization; if the goal is to increase the production of food and vegetable raw materials, it will be necessary to clear forests, resulting in less space for carbon sequestration. If the goal is to plant biofuels, there will be less space for food and forests. There are many choices along the way.

As resources become scarcer, countries, cities, and communities most efficient in the consumption of energy will be successful in promoting better quality of life.

Reducing the ecological footprint is critical towards this goal. Unequal distribution of natural resources by nations means our perceptions of resource depletion are not fully visible. What is scarce in one region can be abundant in another. What really counts are the global levels and the rate of increase in consumption.

Another concept using similar images is the water footprint, established by the Dutch scientist Arjen Hoekstra, from the University of Twente in the Netherlands.[23]

Water is not a biologically produced material, rather it is a resource that is part of the biosphere, existing in constant quantities on Earth. Biological cycles recycle and move water, but they do not produce it. Since it is not included in the calculation of the ecological footprint, Hoekstra felt water deserved its own indicator. For human needs, what matters is the availability of clean water.

The water footprint measures the demand of water when used in the production of material goods. The examples are impressive: in average, in order to produce a cotton shirt, 2,700 litres of water are spent; 1 kilo of beef requires 15,500 litres; 1 kilo of soybeans, 1,800 litres. Agricultural and industrial processes still use too much water, with very little efficiency. Fortunately, the water footprint concept is already leaving the universities and becoming a tool for strategic planning in companies, still few at the moment, though. Critics argue that, as the measurements refer to averages, there is no precision in the concept. However, it is a helpful beginning towards perceiving how a resource that has always been considered free is becoming increasingly scarce, tending to become more and more expensive.

Many countries export their water footprint when they sell products requiring intensive use of water. Brazil belongs to this list for being a major exporter of soybeans, beef, cotton and coffee, which puts water resources from the exporter's regions under pressure.

THE MOST PRECIOUS ASSET

If the 20th century was the century of oil, the 21st will be, as we have seen, the century of water.[24] The challenge having the most impact on us is how we can ensure quality without destroying ecosystems or watersheds.

There exists the chance that we will not have water in the quantity needed for consumption, especially if we continue using wasteful practices and polluting our water resources. Water is becoming increasingly scarce, and geopolitical conflicts involving water are already being projected.

The use of fresh water from our ecosystems is beyond the point of sustainability, which means we take more than the ability to regenerate a supply of clean water can keep up with. The greatest impacts on water resources include excessive withdrawal from springs, river pollution, and human interventions such as canals, dams, locks, and waterways. Climate change tends to exacerbate the problem.

The report from UNESCO's World Water Assessment Programme, "Water in a Changing World Report 2009," estimates that in 2030, 47% of the world population will be living in areas under hydric stress (shortages or lack of water). Five billion people will be without adequate sanitation facilities—or 67% of the world. The quality of water directly influences food security: fishes are the primary animal protein consumed in many countries.

Dams and dikes have been built in most large rivers around the world. Of the 177 rivers longer than a thousand kilometres, only 64 flow freely with no barriers, according to the "Living Planet Report 2010" by the WWF.[25] Infrastructure built in rivers brings benefits, but at the same time impacts ecosystems. Dams alter the regime of rivers, the speed, and the amount of water pouring downstream. Large dams can block ecological connections between habitats upstream and those downstream, like fishes in need of migrating from the lowest to the highest part of the river in order to reproduce.

The extraction of water has led to drought in some of the planet's great rivers. The Yellow River in China, the Murray in Australia, and the Rio Grande on the U.S. border with Mexico are already dry at some points of their routes. In China, there are large-scale transfers of water from south to north. In Brazil, the project of water diversion from the São Francisco River raises controversy as well as social, environmental, and economic disputes.

Pollution of waterways is a persistent problem, and also an issue of moral conscience. Those who live upstream collect clean water and return contaminated water to the user downstream. Watersheds are the circulatory system of the Earth and, like our veins and arteries, carry the vital fluid throughout the body of the planet. We owe the rivers much more respect.

ADAPTING BOATS TO RIVERS, NOT RIVERS TO BOATS

One of the good fights I have engaged in happened in 1997. I was invited to contribute to a publication about the Paraguay-Paraná Waterway Project.[26] The study was a collaboration with the EDF (Environmental Defense Fund), an American non-governmental organization which, at the time, was associated with CEBRAC (Center for Reference and Cultural Support Foundation). The aim of the study was to support public policies.

The government of Paraguay and the contractors had obvious interest in the work. Free navigation of the rivers of Rio de La Plata Basin has always been one of the goals of the foreign policy of the Brazilian state and also of countries wishing to establish trade in the inner regions of the continent, especially England. The South American integration basin had been the basin of the disputes that culminated in the Paraguayan War, basically a naval war, in which the Navy of the Triple Alliance (Brazil, Argentina and Uruguay) sailed upstream in the Plata river to Paraguay and to the state of Mato Grosso, in Brazil.

In the case of the project for the Paraguay-Paraná Waterway, the issue became dramatic and highly controversial. It aimed to transform the Paraguay and Paraná rivers into a navigation canal to transport agricultural and industrial production to ports on the coast.

The project started development in the 1980s, through negotiations between Argentina, Uruguay, Paraguay, Bolivia, and Brazil. The feasibility studies were financed by the Inter-American Development Bank and the work developed by an intergovernmental waterway committee. The project proposed a waterway providing dislocation from the city of

Cáceres, in the state of Mato Grosso (Brazil), to the port of Nueva Palmira in the far south of Uruguay. This uninterrupted waterway would be more than three thousand kilometres long and allow day and night navigation throughout the year. Of its total length, one third would be in Brazil.

In an article published in the newspaper *Folha de S. Paulo,* in March 1997, I warned that

the main controversies were of two natures. Firstly, direct and indirect environmental impacts were unknown, the studies made not being satisfactory. Secondly, the real need for those sustainable *development works by the large region it would service was discussed.*[27]

Several stretches of those rivers would receive works of plumbing, drainage, removal of rocks from the bottom, cutting of banks, dredging, and the construction of ports and locks. There was little knowledge on the dynamic functioning of the hydrological regime of the Paraguay River Basin, which depends on rainfall, the geomorphology of its microregions, and on interactions between flora and fauna.

The water level of the Paraguay river critically determines the portion of floodplains, with just a few changes in the morphological structure of the river being enough to prevent large areas from being flooded. This causes great losses in biodiversity. Studies already showed that a lowering of the Paraguay River quota would seriously affect the water regime of the Pantanal.

The mathematical model on which the project was based was incomplete because it lacked much preliminary data and comprehensive studies on the complex dynamics. We could foresee some inevitable consequences: an increase in water pollution (with oil, grease, and spills), changes to flooded areas, changes in water flow, increased sediment loads in rivers, erosion of land, attraction of population, increased urbanization and cultivated areas, and impacts on fishing due to changes in aquatic vegetation.

Brazil would benefit the least and would bear the largest environmental and social impact. In short: everything we do not want when trying to build a sustainable future. The loss of biodiversity in the Pantanal would be the most catastrophic consequence.

The Pantanal serves as a system for the transportation of diffused water. All dynamic processes depend on water to keep their high productivity and richness. And the whole of the Pantanal's life is a result of this balance. Contributing to prevent the implementation of the Paraguay-Paraná Waterway was one of the great moments for our team.

Any change to the morphology of the Paraguay river would be reckless, and the defenders of this project were the usual suspects: international and national companies linked to the provision of barges and large-scale works of the dredging and straightening of the river, well known destroyers who, under the aegis of development and production of wealth, are nothing more than predators at any cost. The intention was to adjust the river to the boats and not the boats to the river.

Amid the negotiations, I was invited to Paraguay, where, in a highly caricatured scenario, I was greeted with celebration. There was also a group of Americans who had built barges designed for the Mississippi River; those barges were no longer of any use, and they wanted to thrust them upon us. There were a conjunction of interests at stake, but we managed to get to the IDB and ultimately prevented the project from going ahead.

I have deep affection for the Pantanal. Unlike the dense forests, the Pantanal openly exposes all its diversity. The water slowly flows to the plain, and it is that journey which guarantees the maintenance of life in its most exuberant forms.

When the Paraguay River begins to recede, the ditches—"currichos," in the language of the Pantanal—send their water to the river. This water is full of nutrients which can feed all species. The animals appearing are fishes, birds, alligators, capybaras and jaguars. This whole complex cycle of life is beautiful for us. But we are the ones who depend on it.

FORESTS FROM THE SEA

Life began in the oceans. They are not only immense masses of water; they are more like a cake with different layers, each a different temperature and salinity. When these layers move they are like huge rivers that flow in different directions, forming the ocean currents.

The interactions between these different layers, and between the oceans as a whole and the atmosphere, are one of the engines of global climate. By polluting the air and the seas, we interfere with complex mechanisms that we have just begun to understand.

Marine ecosystems are among the most productive in the world, especially those in the waters of the continental shelves which sustain much of oceanic fishing. In them occurs a phenomenon called upwelling, which returns nutrients deposited on the seabed to the surface. Phytoplankton and algae can use the sunlight for their photosynthesis, producing organic matter that will be a source of food to the whole food chain.

The microscopic marine animals (zooplankton) eat phytoplankton (microscopic algae) and are the meal of fishes and small crustaceans, which are then eaten by bigger fishes, up to sharks and whales. The micro feeds the macro.

The most biodiverse of all marine ecosystems is near coral reefs. Coral colonies can grow and create very large structures, which eventually serve as food and shelter for fishes and marine animals of all kinds. It is estimated that there are between one and two million species living in coral reefs throughout the world. From this amount, only about 10% have been identified and described.[28]

For harbouring an extraordinary variety of plants and animals, they are considered the most diverse marine habitat in the world. A single coral reef can harbour at least three thousand animal species. Corals clean the ocean, acting as a filter for impurities and excess of organic substances.

The areas of coral reefs are considered, in terms of biodiversity, equivalent to the rainforests. There is a potential for bioprospecting, and their socioeconomic importance cannot be underestimated.

The environmental goods and services provided by corals are reflected mainly in an abundance of fish, but they are not limited to that: active chemical substances have been discovered which may be the basis for future drugs. Protecting coral reefs is a critical element of social and economic planning. Where corals are preserved, there are more marine resources, more fish and higher touristic potential, which can generate income and employment for coastal communities.

There are constraints. While at the same time a source of biodiversity, coral reefs are fragile ecosystems sensitive to change. They are threatened by factors such as sedimentation, where particles from poor agricultural practices and deforestation are brought through rivers to the sea. Those particles may obscure sunlight, interfering with the algae's photosynthesis. Pollution from industrial effluents and chemical fertilizers also pose a threat because it can induce the excessive proliferation of algae, which leads to a possible shortage of oxygen for marine animals.

The biggest threat to coral reefs worldwide is climate change. The temperature rise of the oceans disturbs the delicate balance of this ecosystem, which is not adapted to withstand strong temperature fluctuations.

The sea is the largest carbon sink, with the cost of alterations in its chemical composition. The increase of CO_2 in the atmosphere causes the oceans to absorb much of that gas in their waters. By reacting with water, CO_2 forms carbonic acid: $CO_2 + H_2O \rightarrow H_2CO_3$. A simple and lethal equation to the marine life cycle.

Carbonic acid makes ocean water more acidic, and all organisms then have to adapt to an environment with a lower pH (hydrogen power). This change is happening fast, and many species will not be able to adapt, eventually becoming extinct.

Corals, clams and mussels have shown difficulties in forming their shells and skeletons, which are becoming more white and fragile.

The global community is already mobilizing for the cause. Countries in the region of the Asia-Pacific promoted an initiative to protect 5.7 million square kilometres of the Coral Triangle, the habitat of 76% of known coral species in the world.[29]

There is a passage in the Hebrew Bible, a basic repository of Judeo-Christian faith, in the book of Ecclesiastes, attributed to King Solomon, with verses appropriate to the limits of sustainability:

"What profit hath man of all his labour wherein he laboureth under the sun? One generation passeth away, and another generation cometh; and the earth abideth for ever. The sun also ariseth, and the sun goeth down, and hasteth to his place where he ariseth. The wind goeth toward the south, and turneth about unto the north; it turneth about continually in its circuit, and the wind returneth again to its circuits. All the rivers run into the sea, yet the sea is not full; unto the place whither the rivers go, thither they go again. All things toil to weariness; man cannot utter it, the eye is not satisfied with seeing, nor the ear filled with hearing. That which hath been is that which shall be, and that which hath been done is that which shall be done; and there is nothing new under the sun. Is there a thing whereof it is said: 'See, this is new'? – it hath been already, in the ages which were before us. There is no remembrance of them of former times; neither shall there be any remembrance of them of latter times that are to come, among those that shall come after." (Ecclesiastes, 1, 3-11).[30]

2

Key questions of the Amazon

Today, Brazil is possibly the richest country on the planet, due to the Amazon's diversity and water resources.

No wonder the Amazon is central in all global discussions on climate change. The devastation of the forest regions of Brazil is among the greatest emitters of greenhouse gases in the world.

There are a few myths surrounding the Amazon. The land seems fertile, but it is not always so. Its soils are shallow, holding few nutrients and are deficient in phosphorus. "In the basin of Rio Negro, the main tributary of the Amazon River, one need only dig a few centimetres into the soil to find sand," says Eneas Salati, our technical director at FBDS for eighteen years and a former director of INPA (National Institute of Amazonian Research).

The Amazon seems homogeneous, but it is not. It has over a hundred different ecosystems with variations in topography, climate, soil, flora and fauna.

It is thus not easy to deal with the Amazon. The world's largest rainforest, the largest river basin in water volume, with its huge biodiversity, is largely unknown and its natural wealth has not yet been assigned a value.

Looking at the Brazilian map, we see that the Amazon takes up almost half of our territory, while most of the whole Amazon River basin belongs to Brazil.

The issue of climate change is closely linked to how we deal with the forest. In the Amazon, the physical and chemical characteristics of the soil depend on the forest, as well as the water balance and the climate.

Vegetation varies according to these characteristics, and may change from dense forest to undergrowth in less than a kilometre. The balance of aquatic and terrestrial fauna also depends on the forest, and so does the culture of the indians and the caboclo people. Any interference with the Amazon Rainforest must be done with extreme care.

THE RIVER OF MANY NAMES

The Amazon River has many different names: it starts in Peru as Vilcanota, snakes its way through the forest and changes to Ucaiali, Urubamba and Marañon. In Peru the Quéchuas, a people descended from the Incas, refer to the Amazon River as "the god who speaks," so strong is the roar of water in its cliffs.

The world's largest river by distance, its water flow contributes to about a fifth of the planet's discharge of inland water into the oceans. This volume of water is so great that it interferes with the temperature and salinity of the oceans for hundreds of kilometres, affecting the climate system in the entire region.[1]

The situation of the Amazon Basin is unique on the planet, leading to extraordinary climatic and ecological consequences.

Over its entire distance, the average depression of the Amazon Plain is very low: from the border between Brazil and Peru to the sea, it is at most ninety meters. From Manaus to the river mouth, the level difference is less than twenty meters—almost nothing. The river flows because of the pressure coming from the Andes. The plain is crossed by numerous tributaries and watercourses, many of which deepened its riverbed creating a more varied topography, with ups and downs.

THE FOREST AND THE CLIMATE

The Amazon is in a state of dynamic balance between vegetation and climate. The Amazonian atmosphere holds vast amounts of water vapour, half of it coming from evapotranspiration, and the other half from the Atlantic Ocean. The vapour thus flows from East to West throughout the year.

The forest recycles the water vapour, increasing the time it stays in the region. The more water is available, the more vegetation grows.

When the forest is cut down this balance is broken, with much of the water vapour escaping from the system. The climate becomes dryer, and the amount of rain decreases. Any human interference should take this basic fact into account. The forest is not a simple consequence of the climate, rather the current balance of the climate is shaped by the forest.[2]

Because of its equatorial location, the length of the day and thus the amount of solar energy remains almost constant throughout the year. Therefore, an important characteristic of the Amazonian weather is isothermia. There are only small variations in temperature over the course of the whole year.

High humidity accelerates decomposition processes. Falling leaves and trees and dead animals are rapidly composted by microorganisms into inorganic nutrients, which are again used by the growing vegetation. It is life feeding life.

The Amazon depends on the standing forest. Everything is an interconnected mechanism: soil composition, water and energy balance and, consequently, climate. The Amazon could never be the world's breadbasket because of the low fertility of its soil and the consequent rapid degradation in ecological balance after its destruction.

FBDS' PIONEER STUDY

Our scientific institutions have relatively few researchers dedicating themselves to the Amazon.

During two decades of study from the FBDS, we have carried out more than two hundred environmental projects, some of which were groundbreaking. The economic-ecological zoning of the Amazon was the first, undertaken in 1994, two years after the creation of the Foundation.

The mapping of the Amazon region was done by the IBGE [Brazilian Institute of Geography and Statistics], jointly with other government institutions, but there were many difficulties, both due to administrative problems and to lack of technical data.

The Secretariat of Strategic Studies of the Presidency – at that time headed by Admiral Mário César Flores, with Ambassador Luiz Augusto de Castro Neves as Executive Secretary – invited us to support IBGE in the completion of this economic-ecological zoning. We hired ten different specialists in ten different themes related to the Amazon, such as: water resources, transportation, conservation areas, mining, energy, and extractive reserves. We gathered expertise, giving support and scientific reasoning that served as guides to IBGE. While we were making thematic analyses, IBGE focused on cartographic mapping.

These studies were presented to the Presidency and to the Congress, being the first territorial organization of the Amazon. We helped conclude a process that previously had no end in sight for over two years.

This work created an innovative methodology—used afterwards in several other projects by the Foundation—which was imitated throughout Brazil in both public and private sectors: combining cartographic information containing satellite images with studies about different disciplines related to the physical, biotic and socio-economic environments of the region.

Previously, each of these dimensions was treated as a separate and autonomous factor. From this social, economic and ecological zoning, a methodology was elaborated where it was possible to verify all these components at the same time, integrating their vectors. From there, we had built a radiograph of the region in all its aspects, able to answer many questions about the Amazon.

The methodology was refined from a zoning scale of 1:1,000,000 to 1:250,000; 1:50,000, 1:10,000 and even more detailed, enabling the management of the territory occupation and the measurement of social and environmental impacts.

THE FUTURE OF THE AMAZON IS OUR FUTURE

Deforestation of the Amazon region is clearly a phenomenon of the last forty years. Ever since, Brazil's emissions of greenhouse gases have reached worrying proportions due to these rampant rates of deforestation.

Climate change is the cause, not the effect. The destruction of forests can lead to global climate change: in the case of the Amazon, deforestation changes the water balance, tends to raise the temperature, and decrease rainfall. Knowing how to manage the abundance of water in the Amazon is most likely going to be one of the major challenges for the 21st century.

In 2005, we experienced an unusual phenomenon in the Amazon: a drought, very likely induced by man-made climate change.

At the time, I pointed out the fact that we already had the fundamental tools for strategic planning (SIVAM/SIPAN), as well as technologies for modifying social and environmental policies for Amazonian development.[3]

Scientists already indicated in 2005 that if the anthropic (man-made) deforestation continued, there would be serious changes to the ecological balance of the region and other areas of South America.

The 2010 drought in the Amazon was even greater than that of 2005, which had been the most intense on record. The Rio Negro water levels lowered to the lowest ever recorded. The Amazon is becoming dryer every year, especially at its devastated edges in the states of Mato Grosso and Pará. There is risk of savanization or desertification if this drought trend persists.

During the 2010 drought in the Amazon, the region reversed its role in the carbon cycle: rather than absorbing this greenhouse gas from the atmosphere, the region began to emit CO_2.[4] The most likely cause was the death of large trees, either by fires or through droughts. Destroying the forest means transforming it from a carbon sink into a carbon emitter.

One of the problems caused by climate change is a greater occurrence of extreme events. In the case of the Amazon region, intense droughts can be followed by torrential rainstorms, causing the death of millions of trees. Preventing deforestation is the best way to protect the Amazon biome from the effects of climate change.

FLYING RIVERS

The Amazon Basin exports water vapour, directly affecting the climate of South America and, indirectly, global climate. The amount of water the forest returns to the atmosphere is so large that it causes the so-called "Rios Voadores" ["Flying Rivers"] phenomenon: invisible watercourses, that is, air currents carrying large amounts of water vapour. A term originally proposed by researcher José Marengo from INPE,

"Flying Rivers" is also the name of a scientific research project idealized by Gérard Moss, who has been collecting water vapour samples with the help of airplanes for over ten years.

Salati, a pioneer in the discussion about evapotranspiration in the Amazon and one of the main collaborators in the project, explains that the "rivers are flying masses of air with large amounts of moisture moving through the continents. Deforestation will surely alter two important characteristics of weather conditions: the temperature and the amount of water vapour in the atmosphere."[5]

Flying rivers carry moisture from the Amazon to the Pantanal, to the Centre-South of the country and even to the La Plata Basin. If it were not for this phenomenon of moisture transportation, these regions would have a much dryer climate and hydrologic regime, which would affect life in the other half of Brazil. Agriculture and urban life would be drastically affected by the lack of moisture and the reduced rainfall. The southeast and midwest regions of Brazil owe much of their prosperity to the moisture from the Amazon. The amount of water vapour transported by these masses of air is greater than the flow of the Amazon River.

AMAZONIAN IMPACTS

As a professional duty, I have followed the events happening in Malaysia, in Southeast Asia. I closely observed the attempt of the very same companies who have eradicated the tropical forests there to settle in the Amazon Basin, starting their advances through Suriname. The exploitation of native tropical forests enriches the rich, especially the importers of wood, impoverishes the poor and local people, and devastates national heritage.[6]

In the recent past, the main factor of the destruction of the Amazon was not linked to any immediate economic purpose, but to the legal problem of land tenure.

The military governments considered the occupation of the Amazon a national security priority, with the slogan "occupy to not give away" ["ocupar para não entregar"]. It was believed that an empty forest could be a threat to sovereignty. Thus, a policy to incentivise occupation of the Amazon was created.

Regional geopolitics and territorial occupation gradually changed, followed by illegal appropriations of land (landgrab), highway constructions, improvements in agriculture, mining, industry, hydropower, and oil and gas exploration. Mines were responsible for mercury contamination in the aquatic environment, especially in the 1980s, generating little or no benefit to the population. We know very well the intense social conflicts marking the history of the Amazon.

Large areas of forest have already been transformed into pasture. The herd has been taking over the original forest. Approximately one third of the Brazilian herd, the largest in the world, is from that region.[7] The increase in prices of agricultural commodities such as soybeans and sugar cane has been pushing the farmers into the forest, while the original commodity of the Amazon – biodiversity – is not even taken into account.

The construction of hydroelectric dams in the Amazon is at the centre of debate in the region and has been the focus of attention of prospective studies by the FBDS. We seek to produce a differential and critical equation in order to formulate a frame of reference for hydroelectric projects in the Amazon.

The theoretical potential of hydroelectric generation in Brazil is 234 thousand megawatts, with only 31% of it being used. Most of the unused potential is found in the Amazon region, mainly in the sub-basins of the rivers Madeira, Xingu, Tapajós and Trombetas (90% of the total). Of this potential, only 30,000 megawatts are considered not to have environmental restrictions, since 80% of the Amazon biome is taken up by protected areas, indigenous lands, and priority areas for biodiversity conservation. It is expected that by 2030, all major hydroelectric possibilities, viable from a technical, social and environmental perspective, will already be in use. We are still able to use about 26,000 megawatts from small hydroelectric plants.

The National Energy Plan 2030 provides that generation of electricity must be tripled in twenty years. In this vision there is the expansion of potential use in the Amazon, as in the Centre-South and Northeast regions, this potential is already close to exhaustion.

I will now summarize the emails and phone calls I exchanged in 2011 with José Goldemberg – a teacher to all of us – about this theme. 1) If the power consumption in Brazil doubles until 2030, we will have reached the same level of Mediterranean countries: Portugal, Spain, and Italy. 2) Projections by the Energy Planning Department of the Ministry of Mines and Energy are that power generation will triple, which corresponds to the consumption of colder countries like the United States and Germany. 3) We observe that the National Energy Plan is based on incorrect premises in its prospects and planning.

We are witnessing a clash: on the one hand, strong pressures to build new plants in the Amazon; on the other is the position of environmentalists, who strongly oppose the projects. Belo Monte, on the Xingu River, has become the most emblematic example. The construction of this plant will divert practically the entire flow of the river. With a total capacity of 11,000 megawatts, the plant will however rarely reach that maximum capacity because it will operate in a run-of-river system, restricting its production to the hydrologic regime of the Xingu River. Discharges in that region are uneven throughout the year and, on average, will generate far less power than the maximum possible amount.

Belo Monte goes against the concept of the *triple bottom line*. There is evidence of environmental, social, and even economic damages since not even the financial cost of the project is clear. That plant will be in the heart of the Amazon, more than three thousand kilometres away from the main consumption centres, requiring long electroducts to bring electricity to those places. It will be necessary to flood hundreds of square kilometres of forests and to displace tens of thousands of residents.

It should be noted that, once work has started, there is no way back. The environmental impact reports are incomplete, and there are no clear studies on the safety of energy production. Much of the funding comes from public bodies and companies, leaving the state responsible for balancing financial and environmental losses.

In addition to Belo Monte, there are at least twenty other hydroelectric projects in various regions of the country, almost all in the Amazon.

There is the risk that Belo Monte will give precedent to the use of the enormous resources of the Amazon as a source of life. The energy potential of the Amazon region must be preserved while we develop much more efficient matrices of clean energy than what we have today. According to the projection of the EPE (Energy Research Company), the capacity of wind farms is expected to increase around 300% over this decade. Brazil has a theoretical wind energy potential on the order of 150,000 megawatts, hardly tapped so far.

Drummond, in "Farewell to Seven Falls"[8] ["Adeus a Sete Quedas"], best translated into poetry the anguish of a development model that does not value the natural capital. Drummond was not referring to the Amazon, but to the Itaipu Dam, whose construction meant the flooding of a Brazilian natural wonder, the Seven Falls, on the Paraná River, but the feeling of the poem remains relevant today.

Seven falls passed me by,
and all seven faded away.
The roar of the waterfalls ceases, and with it
the memory of the indians, pulverized,
no longer arouses the slightest chill.
The seven ghosts will join the dead Spaniards,
the dead explorers,
the extinguished fires of Ciudad Real de Guaira.
The seven ghosts of the waters
murdered by the hand of man, owner of the planet.

Here once rumbled voices
of imaginative, fertile nature
in theatrical productions from dreams
offered to men with no contract.
A beauty-in-itself, a fantastic drawing
embodied in foaming waves and airy mists
showed itself, stripped, gave itself
in free coitus to the ecstatic human vision.
All architecture, all engineering
of remote Egyptians and Assyrians
would in vain dare to create such a monument.

And it disintegrates
due to the unpleasant intervention of technocrats.
Here seven visions, seven
liquid sculptures
vanished through the computerized calculations
of a country ceasing to be human
in order to become a chilly corporation, nothing more.

A movement becomes a dam,
agitation becomes an entrepreneurial
silence, of hydroelectric project.
We will offer all comfort
paid-for light and power generated
at the expense of another good without price
or redemption, impoverishing life
in the fierce illusion of enriching it.
Seven herds of water, seven white bulls,
of billions of white bulls integrated,
sink into pond, and into the void
no form will occupy; what remains
from nature except for pain without gesture,
silent reproach
and the curse time will bring?

Come, strange peoples, come, Brazilian
brothers of all faces,
Come and see and keep
not the natural art work
today postcard in colour, melancholic,
but its still roaring spectre
of iridescent pearls of foam and rage,
passing, flying in circles,
among destroyed suspension bridges
and the useless lamentation of things,
rousing no remorse,

no burning and confessed guilt.
("We take responsibility!
We're building a great Brazil!")
And yadda yadda yadda...

Seven falls passed us by,
and we didn't know, ah, we didn't know how to love them,
and all seven were killed,
and all seven disappeared into thin air,
seven ghosts, seven crimes
of the living taking a life
never again to be reborn.

3

Energy and climate change

The short-term perspective ignores that which is no longer a forecast, but a fact which is absolutely crucial for the next twenty or thirty years.[1]

There is no production or consumption without impact. We are experiencing a challenge, that of producing energy with less impact and lower risk. Industrialization and advances in medicine worked so well that the world's population jumped from just over three billion in 1960 to nearly seven billion in 2011. Urban concentrations have been steadily increasing. By 2050, 70% of the population on the planet will be living in cities. This is the panorama that faces us. A considerable portion of the population will need to have access to the energy denied to them today. According to the projections, in the business-as-usual scenario there will still be 1.3 billion people without electricity in 2035, which will require a huge effort in order to meet this additional demand.

EIA (U.S. Energy Information Administration) and IEA (International Energy Agency), linked to the OECD, made overlapping projections that global energy consumption will increase 49% by 2035.

We have yet another challenge: to produce energy without aggravating the environmental problems man has already caused on Earth. A significant change in habits and consumption patterns will be necessary.

ENERGY, INHERENT TO LIFE

Human history can be characterized by the energy matrix adopted by man. The only source of energy for primitive man was his muscles.

Then he began to use fire and the domestication of animals for his own benefit, such as horses, indispensable energy sources for transportation.

With each new discovery and invention, more energy began to be consumed by society, and even more after the Industrial Revolution.

The following data illustrate the evolution of energy consumption:

Primitive man	2.000kcal/day
Hunter	4.000kcal/day
Farmer	12.000kcal/day
Industrial	77.000kcal/day
Technological (post-industrial)	230.000kcal/day[2]

The success of an industrial society, the quality of life for its inhabitants, the growth of the economy and its impacts on the environment are largely determined by the quantity and quality of its energy sources and by the efficiency with which its systems transform potential energy into work and heat.

Earl Cook, an American geologist, in his book *Man, Energy and Society* (1976), described how man got involved with energy cycles, converting energy from natural sources into more desirable forms: from pasture to steak, from wood to heat, from coal to electricity. Cook was already worried about the dangers associated with economic growth and with the fact that this growth had been based on finite and non-renewable reserves of fossil fuels.

What characterizes industrial and post-industrial society is the use of energy capitals, such as the burning of fossil fuels long stored underground. We live from this capital, accumulated over millennia by nature.

Technological man already consumes more than a hundred times more energy than primitive man. Humanity's use of energy increased so much that it changed the face of the Earth. At night, instead of darkness, the planet shows large illuminated areas and spots where there are urban concentrations and great dark areas of deserts, untouched habitats or underdeveloped regions.

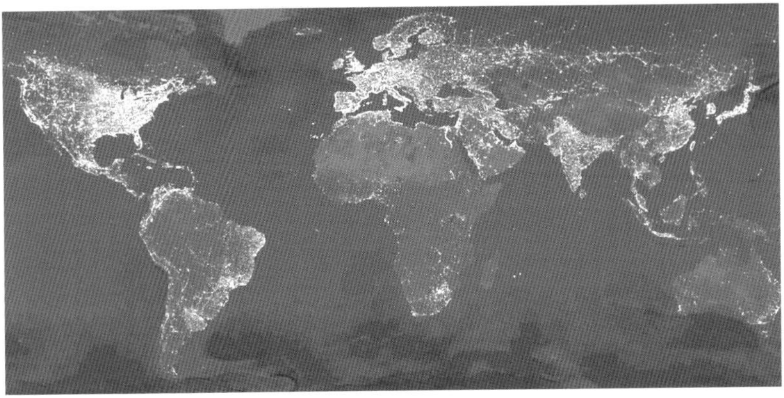

Earth at night – october 5, 2008.[3]

The future of civilization depends mainly on the choices we make in dealing with the energy issue. The models used to make predictions about the economic and climate scenarios are hindered by a significant degree of uncertainty, resulting from estimates of energy consumption increase and population size. The amount of energy *per capita*, and from what source, is still a challenging question for us to answer, and this answer concerns us all.

Several studies in the last years have shown this unsustainability from an environmental, economic, and social viewpoint, of the way energy has been produced and employed. Energy patterns need to change quickly in ten or twenty years if we want to decrease the impact of the use of fossil fuels on Earth's climate dynamics. Without changes in *realpolitik*, the world moves toward an increase in average temperature of up to six degrees by the end of this century, 2100, with catastrophic consequences.

Urgent action is needed, at the global level, towards low carbon technologies.

The imposition of automotive energy matrices, great emitters of greenhouse gases, results in extraordinarily efficient lobbies which do not take sufficiently into account the research and development of new technologies and alternative sources of energy.[4]

As much as public awareness has grown about the problem and, although environmental issues have been occupying more and more pages of the news, we are still far from being able to face the oil-coal-gas complex, driver of the world economy. The solutions must be designed top-to-bottom to undo the knot tightly tied by the fossil-based economy.

BLACK STONE VEINS THROUGH THE MOUNTAINS

Coal was the main source of energy in the beginning of the industrial era, used as a fuel in steam machines, where its thermal energy was converted into mechanical energy. Its heat capacity had already been known for a very long time.

The use of coal for domestic heating was recorded by the Romans in the province of Britannia (today England). When Emperor Claudius conquered the island, in 43 AD, the Romans began to commercialize coal and enjoy new comforts. The Roman Empire bathed in water heated by hypocausts (a kind of heating system) installed in public baths, forts and in the houses of wealthy citizens.

In the Middle Ages, coal was replaced by wood. It was only in the eighteenth century, with the beginning of the Industrial Revolution, that coal returned, with the development of more efficient extraction techniques.

The Chinese already knew coal and its properties and mined it on a large scale, by pre-industrial standards. Its use was widespread in the cities and subject to taxation by the imperial governments. A testimony of this contrast, unknown in Europe but common in medieval China, is found in the writings of the merchant and traveller Marco Polo, who served in the court of the Sino-Mongolian Emperor Kublai Khan, at the end of the thirteenth century.

"... black stone, which they dig out of the mountains, where it runs in veins. When lighted, it burns like charcoal, and retains the fire much better than wood; insomuch that it may be preserved during the night, and in the morning be found still burning. [...] It is true there is no scarcity of wood in the country, but the multitude of inhabitants is so immense, and their stoves and baths, which are continually heating so numerous, that the quantity could not supply the demand; for there is no person who does not frequent the warm bath at least three times in the week[.]"[5]

WHAT THE FUTURISTS SAID

Petroleum (etymologically "oil from stone") is a substance known since Antiquity, found in natural wells in the Middle East and Central Asia. The beginning of the modern oil industry is considered to have begun only in the mid-nineteenth century, with the drilling method invented by Drake in Pennsylvania. A substitute for whale oil, which was becoming increasingly scarce and expensive, the main economic use of oil was in the burning of kerosene for lighting. The development of refining processes and of an efficient distribution system (pipelines) promoted its economic and political rise.

Oil found its *raison d'être* with the development of combustion engine vehicles. Gasoline and diesel fuel had become the major sources of energy for transport, but it was not an easy beginning. In late nineteenth century, different technological matrices competed, and this race would leave its definite mark on the twentieth century. Cars powered by electricity were also being developed, and combustion engines were initially less efficient and harder to start.

The competition between the electric propulsion motor and the combustion engine ended with the invention of the new Ford Model T, in 1909, with low production cost and the change from the heavy and dangerous crank mechanism to electric ignition. The low cost and mass production of the Model T, together with the interests of oil companies, made the noisy and dirty combustion engine the industry standard.

Perhaps oil vehicles best fit the spirit of the *Belle Époque,* proud of the rise of industrial and technological civilization, with its immense machinery emitting smoke. The futurist movement reflected very well the triumph of the oil-automotive form of transportation:

"We declare that the splendour of the world has been enriched by a new beauty: the beauty of speed. A racing automobile with its bonnet adorned with great tubes like serpents with explosive breath... a roaring motor car which seems to run on machine-gun fire, is more beautiful than the Victory of Samothrace. We want to sing the man at the wheel, the ideal axis of which crosses the earth, itself hurled along its orbit."[6]

When man started to burn coal and oil on an industrial scale, it seemed like a good idea, in a *post factum* view. Coal and oil are reserves of organic matter transformed by pressure and temperature underground: whole forests, flora and fauna buried millions of years ago. We are burning fossils, concentrated energy sources which offer the advantage of being easily transported and distributed. But we pay an increasingly high price and have been suffering the effects on global climate caused by growing CO_2 emissions. Around 75% of the global emissions of greenhouse gases come from the burning of fossil fuels.

HOW LONG WILL WE BE ABLE TO BURN FOSSIL FUELS?

The world consumed, in 2007, about 85 million barrels of oil per day. By the end of that year, the world's proven reserves were approximately 1.4 trillion barrels. If the pace of consumption and the size of reserves remain the same, the world will still be able to burn oil in the next forty years. Oil shortage will not force us to replace it with renewable, cleaner sources, but the urgent need to drastically reduce emissions of greenhouse gases will, through political decisions leading to oil's replacement.

The predictions from energy agencies are worrying. A considerable increase in the production and consumption of all fossil fuels is expected, burning the reserves even faster if the current business-as-usual model continues. This increase will require surveys and new technologies in order to find new deposits.

For decades, the countries owning the largest reserves have been the same: Saudi Arabia, Iran, Iraq, Kuwait, Venezuela and the United Arab Emirates. The largest are located in the Middle East, a politically unstable region, with strong social demands. The West has always regarded the Arab countries as oil suppliers, regardless of them being dictatorships or closed monarchic regimes, as long as they continued to supply oil. Producers, distributors and speculators became more and more powerful. In the 1970s, the so-called petrodollars appeared as a result of the financial impact caused by rising oil prices.

We can see, however, opportunities in the almost permanent instability of the global oil market. The higher the oil price, in principle, the more it will encourage the search for new technologies and cleaner energy sources.

We must not forget that, before consumption, exploration, extraction, and transportation involve huge environmental, economic and social risks. We need to be more aware of the impacts caused by possible accidents, and not only of the effects of oil burning.

The oil spill in the Gulf of Mexico in 2010 – after an accident on British Petroleum's Deepwater Horizon platform – was just one in a long series of environmental disasters involving oil exploration. What was the size of this disaster? The answer depends on whom you ask. The BP shares fell by half and ended up being frozen in the capital markets. But this was not even the worst oil spill in history, not even the one with the highest social and environmental damage.

As the attention turned to the Gulf of Mexico, in 2010 the New York Times published a story about the oil exploration in Nigeria.[7] According to the newspaper, the Niger River Delta has been undergoing the equivalent to the oil spill of the accident of the ship *Exxon Valdez* in Alaska in 1989, when 260,000 barrels (official version) or 750,000 barrels (environmental organizations version) of oil were spilled.

Nigeria produced more than two million barrels of oil per day in 2009, and thousands of kilometres of pipelines have been built through fields over the last fifty years. The Niger Delta region contributes to 80% of government revenue, but for decades a silent environmental disaster has been affecting that corner of Africa. Old plants, pipelines, damaged pipes, abandoned equipment, and unprofitable oil fields continuously leaking are common in the region. The local population does not benefit from oil revenues, and life expectancy in this region is even lower than in the rest of the country. The Niger Delta is a very biodiverse environment, an interface between the ecosystems of coastal areas, forests, wetlands, and rivers. There is much resentment from the local population against oil companies, because the fish and shrimps have disappeared from the region. Every year, local environmentalists denounce the situation, but are given little attention.

What about the coal? Next to oil, coal has remained the world's second largest energy source. Current consumption is about 6.7 billion tons per year. World reserves total 826 billion tons. Assuming that the world production is constant, the reserves that already exist are sufficient for 130 years. The largest are in the United States, Russia, China, Australia and India.

If coal, a raw material, reminds us of the early days of industrialization, of dirty factories, the reality is that it is still the basis of many economies. Billions of people still depend on it.

China sits on a sea of coal, a matrix ensuring almost 80% of its electricity generation. The Chinese are building new plants at a fast rate. Unlike natural gas and oil, whose reserves are more limited, coal is still abundant and cheap, which encourages its use. Therefore, the large supply and relatively low price of coal on the international market explain the predominant world consumption of this leader in the emission of greenhouse gases.

A kilowatt/hour of electricity generated from coal costs about 6 US cents, while the same kilowatt/hour from solar energy costs 40 cents.[8] The cheap almost always comes out expensive. This is the kind of *trade-off* we have to analyse and ask ourselves what the real cost of coal is; is it only 6 cents? What would its real price be if we counted degradation to our health and the cost of extreme climate events? The true cost of coal is much higher than the cost of solar energy. This is called internalization of the cost of environmental impacts, which today are paid indirectly by mankind. This reasoning is applicable to all other fossil fuels.

On the production side, coal mining is still a dangerous and unhealthy activity, with high social and environmental costs. In China, in the Shanxi province, there are approximately 1,500 mines, where fires and landslides are frequent.[9] The tunnels of the mines are so low that workers can only move by crouching or crawling. The salary is about 170 dollars a month for this kind of work. In the United States, in Kentucky, coal companies explode the top of entire mountains to more easily reach the ore. In South Africa, acidic mine water from deactivated mines pollutes rivers. In Colombia, mining companies are displacing families to expand the Cerrejon mine, one of the largest in the world.

Natural gas is the third most consumed fuel in the world, reaching today more than three trillion cubic metres per year. The world reserves are 187 trillion cubic metres, which would ensure more than sixty-two years of consumption at current levels.[10] Russia, Iran, and Qatar hold the largest reserves in the world. Among fossil fuels, natural gas has the lowest emission level.

Geopolitics and energy savings go hand in hand. Natural gas has been used as a powerful advantage in foreign policies between Russia and Europe, for instance.[11]

The environment should be treated outside partisan lines and not fall into ideological or sectarian traps. This is a matter of life or death to the extent we depend on the responsible use of energy to ensure life. Will we have to deplete all fossil fuel reserves, or will we be able to make a qualitative leap in the areas of energy, transport and industry?

WHEN THE SOLUTION BECOMES A PROBLEM

Nuclear power accounts for 15% of all electricity generation worldwide. The estimate that nuclear power will contribute to 6% in the reduction of greenhouse gases emissions by 2050 will have to be reviewed. The earthquake followed by a tsunami that struck Japan in March 2011 started a global questioning on the safety and convenience of this energy source. The disaster damaged the reactors of the Fukushima-Dalichi plant, contaminating a large area of this small but populous country with radiation. Thanks to the solidarity and ability of the Japanese people, the disaster did not have more serious consequences.

In recent years, there has been an attempt to rehabilitate nuclear power, in light of possible solutions to diminish emissions associated with energy, since the atomic fission process does not emit greenhouse gases. Its proponents sought to minimize the risks of accidents and the issue of nuclear waste. Chernobyl (1986) seemed very distant in time.

Germany announced the closure of all its nuclear plants by 2022, focusing on the transition to a new energy grid based on efficiency and clean sources like wind, solar, and hydroelectric energy. This political decision came from the popular pressure against nuclear power following the disaster of Fukushima. With this, Germany will further encourage the development of new technologies, reducing their costs. Meanwhile, China is today the country that expands its nuclear power stations the most – of the sixty-six units under construction in the world at the end of 2010, twenty-seven are in China – and there are plans to build thirty more units over the next few years. It is unclear if the accident in Fukushima will change those plans.[12]

According to the latest report by OECD/IEA 2011, Brazil is the fifth largest investor into the research and development of nuclear technology. The country spent a total of 144 million dollars in 2010 in this area. Brazilian plants are located in an important environmental and touristic preservation area in Angra dos Reis, state of Rio de Janeiro, and the infrastructure is too weak to withstand an accident.

CRISES ARE PART OF LIFE CYCLES

What is the future of the production and consumption of energy in the world and the consequent emission of greenhouse gases? A possible, but totally undesirable, scenario is the continuation of the current model, or business-as-usual, in which there would be no significant change in awareness or public policies in relation to the use of energy. In this case, the annual emissions of CO_2 – according to projections made by international energy agencies (IEA and EIA/DOE) – caused by energy would rise from 29.7 billion tons in 2007 to 33.8 billion tons in 2035, an increase of 43%.

No crisis so far has been able to drive the planet away from an extensive use of fossil fuels. But what we are experiencing today is something new. Technologies that can change the energy matrix already exist. What we are seeing are persistent deadlocks in international negotiations. The economy, essentially linked to finance, has led us to a world of markets. It seems clear that we cannot completely give these markets total responsibility for the future we want.

What about China, the greatest carbon emitter, energy consumer, and the country investing most in renewable sources? We are already watching and will continue watching an exponential increase in Chinese energy consumption. In 2000, China consumed half of the energy the United States consumed; in 2009, it had already surpassed it.

China and the United States are the two major energy consumers on the planet. However, in a *per capita* rate, the United States remains first.

China is already the leader in wind and solar photovoltaic energy. Only by the year 2009 did the development of solar panels have such a great impulse that prices fell from 59 US cents a kilowatt-hour to 16 cents.[13] By the end of this decade, China may come to dominate the global production of power equipment. Incentives for clean energy should further strengthen the equipment industry.

In the latest five-year plan, released in March 2011, for the first time energy efficient industries were declared a priority, and investments of 400 billion dollars were announced for environmental protection, twice as much in relation to the previous plan.[14] According to Karl Hallding, an expert in environmental policy for SEI (Stockholm Environment Institute), the Chinese are already thinking about carbon taxation. Despite their large reserves, the Chinese already import coal. How to ensure energy in a country that grows at an average of 10% a year?

China heavily copied the Western market model and depends on it to keep its exports high and produce reserves needed to enable stable domestic policies. Chinese growth is coming to its physical limit: "In China's thousands of years of civilisation, the conflict between humanity and nature has never been as serious as it is today. The depletion, deterioration and exhaustion of resources and the deterioration of the environment have become serious bottlenecks constraining economic and social development", declared Chinese Environment Minister Zhou Shengxian, in June 2011.[15] China still maintains huge poverty, more than 900 million people do not benefit from this model.

Brazil has a clean energy matrix whose main source is hydroelectricity, which is renewable and does not emit carbon. To reduce emissions in Brazil, we must halt deforestation and only then will we be in a privileged position ahead of the major economies of the planet.

THE PLANET BREATHES

The theme of man-made climate change started to come to the attention of the scientific community from studies made in the late 1950s at the observatory of Mauna Loa, an isolated island in Hawaii. These systematic measurements were gradually extended in the 1960s and 1970s to other stations, from the Arctic to Antarctica.

The main scientist of this CO_2 measurement programme, the American Dave Keeling, made two important discoveries: a steady increase of CO_2 concentration and great annual fluctuation of this gas, with minimum numbers in the summer and maximum in the winter in the northern hemisphere. In spring, the biosphere removes CO_2 from the atmosphere, plants start to grow and, when autumn comes, the leaves fall, the cycle is reversed and the carbon in organic matter goes back to the atmosphere as CO2. This is called the planet's "breathing."

The data indicated the importance of biological processes, especially photosynthesis, in the flow of carbon. The following chart indicates what became known in the history of climate change as the Keeling curve. These measurements were one of the instruments relevant to the first IPCC report in 1990, and the scientific basis for the Climate Convention, signed in 1992.

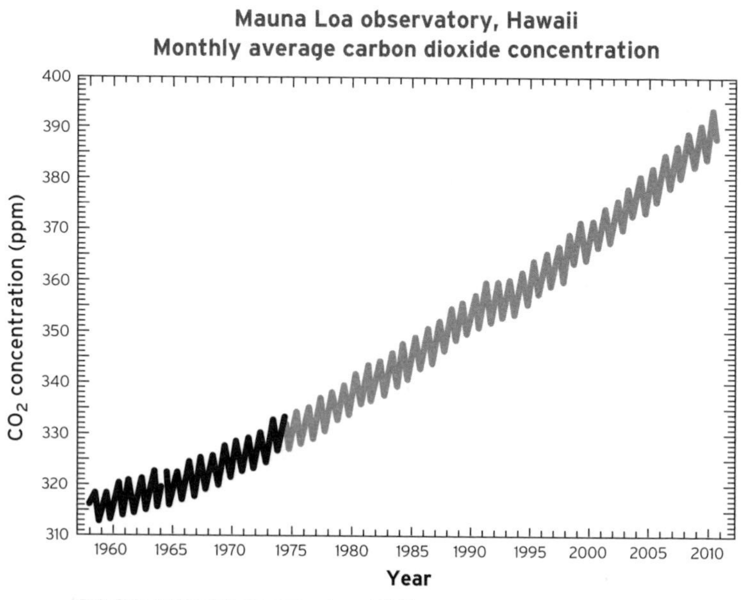

Data from Scripps CO_2 Program – August 2010.

The concentration of this gas in the atmosphere since the Industrial Revolution has gone from 280ppm to almost 380ppm, and keeps on growing. The problem of this increase is that CO_2 is characterized as a greenhouse gas, i.e., it has high capacity to retain energy, which would be partly emitted into space in the form of infrared radiation (heat) – a heat which remains in the atmosphere. This simple fact already causes huge damages which may intensify in the future. Therefore, energy that is apparently cheap today has expensive consequences.

To prevent further global warming, the scientific consensus has established a maximum increase in global temperature of 2°C. But the reality is that we do not know the exact consequences of this level of increase.

TWIN CHALLENGES

What are the ways to limit the concentration of greenhouse gases in the atmosphere? We will have to start with a strong political will, coupled with new technologies and energy efficiency gains. We must use all available tools. Energy efficiency is a factor of increasing awareness throughout the production chain.

The more countries participate in mitigation and the more economic sectors involved, it will be all the more cheaper and efficient to reverse the upward trend of global emissions. Governments and companies are co-responsible in this new architecture of the energy matrix.

One of the most efficient policies to reduce GHG emissions is carbon taxation. Polluting is expensive, and not only to the environment. This cost needs to be significant for emitters. Investments in technology and low carbon energy sources have increased in recent years but are still small and fragmented in face of the persistence of conventional practices. It is imperative that more aggressive energy policies be adopted, eliminating subsidies for fossil fuels and providing incentives for cleaner and efficient forms of energy.

More recent studies and observations of climate change show that the projection published in the last IPCC report in 2007 underestimated the problem. The climate is changing faster than scientists had predicted. The strongest evidence is the melting of "eternal" snows, especially on the peaks of high mountains such as the Andes in South America, and the Kilimanjaro, in Africa. These are evidence to the naked eye. The goal of 50% in 2050, that is, 50% reduction of CO_2 emissions by the year 2050 in relation to the levels of 2000, may be insufficient to prevent dangerous climate changes.

In July 1997, I was invited to participate in a seminar whose main theme was Renewable Energy and Energy Efficiency for Sustainable Development. It was already clear that we were crossing the threshold of safety and responsibility in the energy issue. Here's my story:

The limit of the availability of natural resources has been making humanity rethink the concepts of development considering a system for the use of technology and natural resources without limitations and without considering the consequences that this same system causes in man's house. (...) The inadequacy of the political, economic and social models is definitive, in view of the need for environmental sustainability on the one hand, and social demands on the other. We live on a geoid loose in the universe, whose average diameter is 12,000 km. Our whole livelihood and capacity of life comes from a merely 10km thick layer surrounding this geoid. The fragility of the atmosphere has been increasingly attacked by uncontrollable emanations, whose absorption capacity by the planet has long been exhausted.[16]

Climate change and energy security are the twin challenges that will shape the future of humanity in the long term.

Developing countries represent at the moment little more than half of the global demand for energy, but this share will increase to 84% by 2035, as can be seen in the following figure.

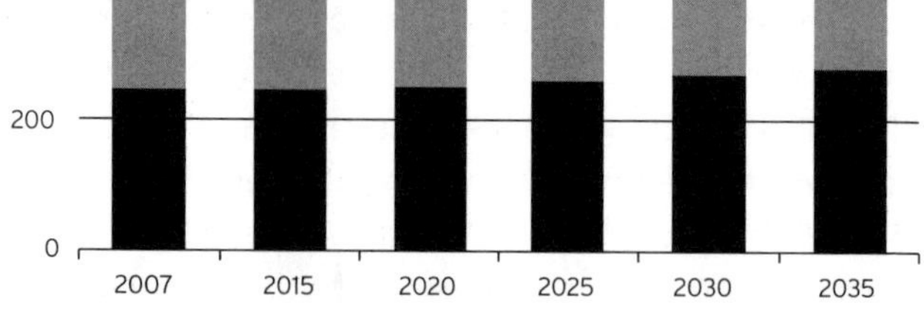

World energy consumption

Source: International Energy Outlook 2010, U.S. Energy Information Administration.

The group of BRIC countries (Brazil, Russia, India and China) accounted for approximately the same amount of greenhouse gases emitted annually by OECD countries until 2005. The strong economic growth particularly in China and India, economies based on fossil fuel reserves, explains the situation and predicts a dramatic increase in greenhouse gas emissions in developing countries over the next few years. Thus, there will be a reversal from the twentieth century: developing countries will respond with twice the annual emissions of developed countries.

It must be noted, however, that the energy consumption and *per capita* emissions of developing countries remain well below those corresponding to industrialized countries (OECD). When comparing cumulative historical emissions, the responsibility of industrialized countries is unquestionable. In 2000, the accumulated emission of industrialized countries was 800 billion tons of CO_2-eq against 200 from developing countries; for the year 2030, it is predicted that the historical emissions will be 1,240 and 685 billion tons of CO_2-eq, respectively.

According to WRI's report in 2006 (World Resources Institute), the United States have the largest participation in the historical cumulative emissions in the period of 1850-2002, with 29.3% of the world's total. The European Union comes second, with 26.5%, Russia with 8.1%, China with 7.6% and Germany with 7.3%. In the period in question, Brazil contributed with 0.8% of total global emissions.

This is a point which has resulted in serious disagreements between countries in the discussion over how to solve the issue of climate change. If it is true that developed countries have to change their consumption patterns, be more efficient in their use of energy and give preference to clean energy sources, developing countries also must share in this effort.

Climate change affects everyone, and its impacts will certainly be more dramatic for the poorest countries. The contributions to the reduction of emissions have to be shared by all countries, although they can be differentiated according to their respective historical responsibilities.

The world must break the link between economic growth and emissions of greenhouse gases through a new green economy, based on low carbon technologies. Reducing the emissions of developed countries in isolation, even if drastically, will not suffice to maintain the elevation of the earth's temperature below 2°C over this century.

MORE ENERGY... FOR WHOM?

In the coming decades an increase in participation of the transportation sector is expected for the total energy consumed globally. The growth of urban population and household income will contribute to the increase in the number of cars *per capita* and in air travel, especially in developing countries. In the world we live in, few win and many pay for the damages.

If we take into account population growth and the increase in energy production, we come to the conclusion that those who already have access will increase their consumption, while those without will remain excluded. In the field of energy, as well as in the economic field, there is serious social disparity.

Today, 20% of the global population – 1.4 billion people – live with no electricity. 40% of the global population – 2.7 billion people – still use traditional biomass (wood) to cook and heat their homes. To provide universal access to energy, the IEA estimates that 36 billion dollars would be necessary per year until 2030, which corresponds to only 3% of the overall investment in energy.

Energy security and food safety go together. The price and availability of food is constrained by the energy supply. The World Bank estimates that currently an increase of 10% in the price of oil is associated with a 2.7% increase in food prices, for three reasons.[17] Firstly, higher fuel prices encourage biofuel plantations (corn, vegetable oils, sugar cane), reducing the area for food production. Secondly, energy causes impacts on the prices of agricultural inputs such as fertilizers, irrigation and pesticides. Finally, transport and distribution of food to markets depend directly on oil prices. In addition to energy, agriculture depends on water resources and on a functional environment. However, climate change, caused mainly by emissions of greenhouse gases from energy production, threatens water resources and the health of ecosystems.

PERVERSE SUBSIDIES: WHAT IS THE REAL PRICE OF ENERGY?

"The days of cheap energy are over" – World Energy Outlook 2009 report.

Energy has never been cheap – economic policies and the practice of not pricing environmental costs led us to think that it was. The truth is that the culture of abundance and energy waste will have to be reviewed.

Excessive use of fossil fuels is largely stimulated by subsidies – financial support, direct or indirect, provided by governments to a particular sector. Artificially low prices encourage energy waste.

The energy sector has a myriad of subsidies, which ends up distorting the real price of the commodity. In the case of a scarce resource, such as energy, it ultimately stimulates consumption. By subsidizing energy demand, governments are able to inflate the rates of economic activity. The result is lower prices for consumers, which tend to result in more emissions of greenhouse gases.

A survey carried out by the World Bank[18], OECD, and International Energy Agency identified thirty-seven countries responsible for 95% of the consumption of subsidized fossil fuels. In 2008, the country that most subsidized resources for the oil economy was Iran, with 101 billion dollars, the equivalent to about a third of the national budget. Subsidies discourage efficiency improvements and drain resources from other sectors.

But it is not only developing countries that practice gasoline populism; both the United States and the European Union countries use this instrument of economic policy. The mechanisms are complex and almost hidden, transferring income from taxpayers to the companies in question.

The mechanisms that stimulate consumption are basically the same worldwide: direct financial transfers, tax reduction and internal control of fuel prices.

On the production side, governments intervene in markets transferring funds directly to companies, sharing business risks, undervaluing public goods and services, lending capital at lower interest rates than the market, or becoming partners in enterprises. When governments subsidize production, the natural tendency is an increase in supply. Since consumption is also subsidized, a vicious circle sets in.

In practice, with subsidies governments signal the abundance and low price of a resource which is in reality scarce and expensive. And it should be even more expensive if we took into consideration the collective damage emissions are causing. Companies benefit, parts of the population benefit, but global society as a whole and the environment lose out a great deal.

Often, subsidies carry the political intention to benefit the poorest of the population. This certainly happens, but is very far from being the rule. In Indonesia, most of the wealthiest families absorb 70% of energy subsidies, while the poor are left with only 15% of them.

Subsidies can be wicked when a wrongheaded political decision is behind them. If, instead of fossil fuels, governments started to fund and subsidize renewable energy sources, the transition to a new economic model would be greatly facilitated.

Seemingly good intentions, coupled with poor public policies, may create the worst of all worlds. With so many disadvantages and negative effects, one has to wonder why subsidies are still strongly used. They also represent, visibly or invisibly, an investment in a certain way of life – which wastes energy, favours individual transportation, is contrary to the scarcity of resources, which prefers conformism to a system already established instead of new models.

The adjustment of prices after the gradual withdrawal of subsidies on fossil fuels and after internalization of environmental costs would promote a more efficient use of energy, and would be capable of encouraging migration to cleaner sources. According to this analysis, the removal of subsidies would reduce CO_2 emissions, but additional measures would be needed. A new model should be the driving force of a new economic cycle we call the green economy.

In an article for *Jornal do Brasil*, in March 2001, I observed that

the price of oil and its derivatives hides a large subsidy, equivalent to the cost of the absorption of the carbon emitted and its consequent environmental impact. Moreover, a quick solution to problems arising from the greenhouse effect and carbon emissions is still distant.

Ten years later, we see a decrease in the distance between problem and solution, but also a persistent lack of political will to change the landscape.

SEQUESTRATING THE PROBLEM

For some time, countries whose economies and energy security are based on the intensive use of fossil fuels will keep on consuming them even if climate policies require the reduction of their use. Still under development, the technology of carbon capture and storage (CCS) will be able to attenuate the situation if this technology sufficiently demonstrates its feasibility.

The process of CCS involves the removal of carbon dioxide – CO_2 – from the emissions of stationary industrial sources, such as power plants, cement factories, and steel mills. Once removed, the gas is compressed and then transported to underground natural deposits.

In 2009, FBDS promoted, together with Shell Brazil, the first Brazilian Seminar on CCS. If oil companies are part of the problem, they must also be part of the solution.

In discussions during the FBDS/Shell seminar, we started from initial studies on this new technology in an attempt to foster global dialogue between authorities, scientists and companies. We wanted to evaluate, at that time, if CCS could be included as one of the solutions allowed by the CDM (Clean Development Mechanism). We discussed the potential technical challenges and implementation costs of the technology and how to make its use feasible in conjunction with traditional ways of generating energy, reducing environmental impact. In this sense, CCS has been studied and experimented in recent years by oil and coal companies seeking alternative ways to produce energy while improving their public image.

The concept of CCS is relatively simple and follows three steps: capture, transport, and storage of CO_2. The transport can be done by compressing the gas and moving it through pipelines, trucks, or ships. Pipelines are obviously the most economical option. For storage, the idea is to use degraded oil and gas deposits or saline aquifers, where CO_2 could be converted into solidified calcium carbonate. At the current stage of development, CCS consumes a lot of energy, is still very expensive and needs to gain confidence in regards to its environmental sustainability.

Despite the doubts, there is some optimism in CCS from the fact that it can be made compatible with the current technological structures and be a bridge until other forms of cleaner energy become feasible. For countries with intensive use of coal and fuel oil, such as China, the United States, and India, it is a solution that can reduce the cost of transition to a decarbonised energy matrix.

The greatest danger of CCS, however, is that it legitimises the use of coal on a large scale, distorting the main goal which is to gain time in the transition from fossil fuels to cleaner sources.

During the FBDS/Shell seminar, I drew attention to the fact that

we are experiencing systemic crises of the current economic, social and environmental models. We are facing the danger of thinking that the solution is in the past and not in the future. In a globalised world, information technology makes us experience the same problems, in all parts of the planet. Dealing with global issues such as climate change in a localized form, under national parameters, will not work. For this reason, we defend the use of all technologies within our reach and that may cooperate with this problem, which is experienced by everyone.[19]

It is vital to face the problem of fossil fuels. By burning fossil organic matter which was under the surface of the earth, man created an imbalance in the physiology of the planet. Old carbon deposits, which geological processes buried deep beneath the Earth, are being released directly into the atmosphere. Not even the best science completely understands the consequences of interfering with the sophisticated mechanisms of climate regulation.

HOW TO UNTIE THE KNOT?

The decisions about what path to take on climate change depends on a common effort.

As has been pointed out by several studies ("Stern Review," 2006; "UNEP's Towards a Green Economy," 2011), the sooner decisions in favour of a low carbon economy are taken, the less costly it will be for the world economy. They show that economic growth and environmental sustainability are not incompatible goals; on the contrary, they can be reconciled.

We need to stabilize emission levels by 2020 under penalty of missing the opportunity to achieve the goal of 80% reduction of GHG emissions by 2050. The more those decisions are postponed, the higher the costs of this process.

I have been defending the thesis that, to cut emissions to the required levels, it is necessary first of all to begin eliminating all form of subsidies for fossil fuels. It is essential to also find a system to incorporate the costs of the externalities of fossil-produced energy, that is, to take into consideration external effects from emissions and residues of fossil combustion on human health and ecosystems. Moreover, it is necessary to penalize emissions of greenhouse gases by setting an appropriate carbon price. The IEA estimates that in 2035, a tonne of emitted carbon may cost between 90 and 120 US dollars. Pricing carbon will serve to faster stimulate a shift in the energy matrix. Emissions from electricity generation, for example, would be immediately reduced.

Reaching the goal of reducing emissions by 50% in 2050 will require, for example, to generate twice as much electricity through renewable sources, compared to 2005 levels.

International agencies have sought to show all possible ways to keep the rise in temperature under 2°C at the lowest possible cost. Climate change will not be mitigated with a single technology. There will rather be the need to adopt a variety of low-carbon technologies, which until now have not been sufficiently adopted, or may be developed for implementation at the lowest cost.

The International Energy Agency envisions two scenarios: the first, if nothing is done (business-as-usual situation), annual emissions can reach 57 billion tons in 2050. In the other, a positive scenario, annual emissions related to energy would be reduced from 28 billion tons to 14 billion in 2050 through the implementation of low-carbon technologies, even with an increase in global energy consumption.

The realization of the desirable scenario would further ensure other benefits, like energy security, by reducing dependency on fossil fuels, and improving public health by reducing air pollution.

Renewable energy would then account for almost half (48%) of electricity generation in the world by 2050. Today, it is only 18%.

The main source of renewable energy is solar energy – from which all other forms of energy derive in some way or another. The technical potential supply of solar energy available in the world is up to 800 times the global demand per year. That is, the Sun is a huge source of energy, but humanity has not yet been able to fully utilize it for its own benefit.

It is true that the use of solar technologies has been growing fast in recent years; for example, between 2005 and 2010, there was a 50% increase. Water heating through solar panels is already economically viable, needing just promotion through more information and funding. Photovoltaic panels, too, have had remarkable price reductions and will become competitive not only for special and remote uses, but for installation integrated with the power grid.

Solar concentrators have been used in different parts of the world, especially in desert areas, and are a great bet for the future. I am confident that solar thermal technology will emerge from a relatively marginal position in the hierarchy of renewable sources of energy to achieve a substantial status among the current leaders in the market, such as hydro and wind power.

Israel was one of the pioneering countries in the use of solar heating in the 1980s, requiring that all new residential buildings were equipped with solar collectors. Today, solar thermal systems are an established technology in the Israeli market of water heating, without any governmental incentive.[20] Since the 1990s I have been seeking to encourage the installation in Brazil of a pilot concentrated solar power plant with technology from the Weizmann Institute of Science, in Israel. Unfortunately, although basic studies have been done defining installation requirements, we could not go ahead and enable greater initiative for lack of official commitment to encompass the idea.

Regarding other renewable energy sources – hydro, wind, and biomass – we can illustrate their importance by the estimated natural reserves and availability to humanity. Hydraulic potential, technically available for years, could supply about 80% of the annual demand of electricity in the world. Wind potential is 1.5 to 9.5 times the same demand, while biomass potential could supply up to eight times the current electricity demand.

Hydropower is the cheapest source of electricity. Recently, new hydroelectric plants have faced resistance because of the huge investments required and because of environmental and social impacts in more sensitive areas. Current installed hydroelectric capacity in the world is around 850,000 megawatts, 11% of which is in Brazil. It is estimated that generation capacity should be at least doubled in relation to the current levels.

The discussion on the expansion of Brazilian hydroelectric capacity is, however, highly emotional, provoking heated debate. Most of what remains to be explored is in the Amazon, and therefore establishes dilemmas and challenges typical of the economic exploration of the great tropical forest in South America, as we have previously seen. Using other forms of renewable energy seems to be a much more efficient way than destroying our forests.

The increase in end-use efficiency of fuels and electricity can also contribute to the reduction of emissions. Policies encouraging energy efficiency should cover all end-use segments – buildings, lighting, appliances, transport and industry.

In the case of architecture, we will need to invest in new ideas for projects focusing on natural lighting and ventilation, incorporating water heating via solar energy and other measures. Design codes and certifications for buildings and civil work are effective instruments for promoting efficiency.

Lighting, for example, accounts for 20% of all world electricity consumption. Until recently, incandescent lamps reigned supreme. The much more efficient CFLs (compact fluorescent lamps), will now dominate the market. The European Union, Japan, and Brazil already have programs for phasing out incandescent bulbs in their markets, making room for the most efficient ones. In addition to the CFLs, other efficient lamps, like LEDs (light emitting diodes), will help reduce consumption.

The use of household appliances is expected to grow sharply in coming years, with a portion of the population so far in the margins of the economy gaining access to these products. The continuing effort on technological improvement should be encouraged as well as the expansion and strengthening of programmes for labelling and standardizing these products.

The transportation sector accounts for 23% of CO_2 emissions related to energy worldwide. There are good perspectives to reduce the use of fuel and cut CO_2 emissions from light passenger vehicles, either through increased efficiency of vehicles with internal combustion engines, or by greater diffusion in the market of hybrid and electric cars with batteries or fuel cells (hydrogen).

In the case of Brazil, efficiency gains in transport will come less from technology and more from infrastructure, as well as from the benefits of using more appropriate logistics and modals. According to a document for public consultation of the National Plan for Energy Efficiency (MME/2010), in Brazil the consumption of fuel to carry a thousand tons of cargo per kilometre is 96 litres on the highways. Consumption in the United States is 15 litres. This shows the extremely low efficiency of our most frequently used modal – road transport – either because of the low quality in infrastructure or the age of the national fleet of trucks. In addition to the precariousness of the highways, 44% of the fleet of trucks has been on the road for more than twenty years and 30%, for more than thirty. The older the vehicle, the more fuel it uses and the higher its emissions. This inefficiency causes not only environmental impacts, but also impacts on the "Custo Brasil" [the high cost of doing business in Brazil], making the country lose competitiveness in international markets.

Second generation biofuels may have a prominent role in the future, such as cellulosic ethanol and biodiesel derived from microalgae. The system of microalgae production still has a high implementation cost, but brings several advantages: CO_2 is trapped by the algae, and their biomass is used as a renewable fuel, directly replacing fossil fuels. Currently, its economic viability is uncertain and will require development efforts in the long term.

It is estimated that hydrogen technology will only be fully viable by 2030 and may significantly contribute to the stabilization of the concentration of greenhouse gases to acceptable levels. But for this to occur, we will need to cheapen the production, transport, and storage of hydrogen. Hydrogen fuel cells must be more durable and reliable, as security is still an issue in the use of hydrogen. We must adopt standards to ensure the safe dissemination of this new technology.

In industry, it will be necessary to adopt on a large scale the technologies currently available, which could be improved and cheapened, as well as develop and implement a set of new technologies, including CCS. If the electricity sector were decarbonised, it would provide new opportunities to the industry by reducing the intensity of CO_2, enabling the electrification of the industrial processes.

The technological substitution of the current installed energy infrastructure will be an unprecedented revolution. It can be a window of opportunity for new forms of development and social inclusion. The energy technology revolution is at hand. The initial costs are high, but will be largely compensated later on, because the benefits will be much greater.

In the end, the ones who invest in an intelligent future will win. We owe this settling of accounts to posterity. Changing the current situation involves an analysis of what we want as a collectivity.

4

Political Model

"My boy, give advice only if you get half on cash." Among my father's teachings, I hold on to some that are very useful in guiding political candidates and their courts.

Politicians in general are unaware of the future as they deal with immediate and sectarian crises. There is no citizen, politicized or not, who does not have a good idea of what a country needs. However, one does not see politicians who have long-term agendas protecting for future generations the heritage of access to the remaining finite natural resources today at our disposal. Vital resources will only last into the future by means of environmentally responsible and socially just projects. I have been fighting for decades for this proposal, which would be "half on cash," ever since the days when sustainability was considered something exotic, from a land of utopian dreamers.

For the first time in history, there is a crucial problem common to all mankind. However, we are experiencing a paradox, a blurred reality. National powers are inadequate to deal with global problems, such as environmental crises, as strongly connected to an energy matrix based on fossil fuels as they are to a lack of proper regulations for an economy of global trading. While financial markets are rational, nationalist feelings are emotional.

"Model" has become one of the most widespread jargon words in the scientific community and the social sciences. What basically interests me when I refer to a model, be it political, economic, or social, is the concept of interaction between these sectors and the final, global form of organization they may take. The concept of model refers here, more precisely, to the structure of several vectors crossing the web of goods, services, and values within which we live.

I like very much Isaiah Berlin's considerations about nationalist sentiments. Berlin, as a Jew who emigrated from Latvia to England, experienced different models of governance during his lifetime. Berlin and I have common roots, since my father's family had to leave Lithuania, in the Baltic region, due to religious and ethnic persecutions. Reading Isaiah Berlin really helped me understand recent political history.

I met him a few times, always in Jerusalem. We had a great mutual friend, Teddy Kollek, mayor of the city for almost thirty years. Walking in Jerusalem, we exchanged many ideas about the State of Israel and Judaism. At Kollek's birthday parties, Berlin and I were the ones responsible for making the speeches. I remember him as an Englishman who was not an Englishman, a very structured thinker with a great capacity for expression. For me, the most important of his works was his last book, *The Crooked Timber of Humanity,* which helped me reflect on globalization and nationalism.

Berlin[1] draws attention to the sentiment of nationalism, which has led to the dominant political model of our time. We assimilate the concept of the homeland from our school years and end up naturally accepting this notion without much reflection. However, many values and attitudes we accept as natural have a recent history and predominated in political struggles against other ideas and world views.

The notion of tolerance, for example, and concepts of freedom and human rights as they are understood today, were created by Enlightenment philosophers just over two hundred years ago and were established as definite political principles from the end of the Second World War. These ideas were based on universalism, and consider all human beings members of one family.

Another innovative idea emerged during the French Revolution and dominated the nineteenth and twentieth centuries: the Nation, with a capital *N*, reached a sacred status, since the existence of the state – formerly under the domain of absolutist monarchy – became legitimized by the Nation, formed by all its citizens. Supposedly by all citizens. Unlike the Enlightenment ideas, the Nation presupposed equality only for its own citizens, giving them higher status than the rest of mankind.

Berlin distinguishes between the sentiments of nationality and nationalism. The first is linked to the need of belonging, group sentiment, identity with others, with family, city or region. Aristotle already recognized the legitimacy of that sentiment. Nationality is based on the notion of common ancestry, which manifests itself in the collective patrimony: language, customs, traditions, historical memories (real or invented), continuous occupation of a territory – those are the elements which, in principle, justify the existence of a state, from its simpler forms (the Greek city-state, for example) to the modern nation states.

Jewish-French writer Ernest Renan created one of the best definitions of nation. In the historiography of Christianity and Judaism, he occupies a prominent place. Dom Pedro II was a kind of epistolary pupil of Renan. They exchanged letters, and the writer was a kind of confidant of the Emperor on Hebrew and Sanskrit. In French newspapers being read in Rio de Janeiro during imperial times, significant space was given to Renan's articles.[2]

Renan, back in 1882, during a conference at the Sorbonne named "Qu'est-ce qu'une nation" (What is a nation?) summarized very well what it is:

"A nation is therefore a large-scale solidarity, constituted by the feeling of the sacrifices that one has made in the past and of those that one is prepared to make in the future. It presupposes a past; it is summarized, however, in the present by a tangible fact, namely, consent, the clearly expressed desire to continue a common life. A nation's existence is, if you will pardon the metaphor, a daily plebiscite, just as an individual's existence is a perpetual affirmation of life. [...] The nations are not something eternal. They had their beginnings and they will end."[3]

Nationalism goes beyond the collective sentiment of nationality, it is a political force that elevates the interests of a nation to the status of supreme values. Later, it would become a defence of the weak (against colonialists) and a justification of the strong (colonialists); it spread beyond its European origins to the Americas, Asia, and Africa, becoming the political paradigm of our time. We have inherited a tradition that took root 150 to 200 years ago, and it will not be easy to free ourselves from it in order to renew the political model.

There have been many thinkers proposing a universalist ideal. For them, nations were barriers that should be torn down to build a society free from rigid political structures. Nationalism was perceived by many as a transitional movement. These thinkers would be surprised to see that nationalism is still very strong in the twenty-first century. For those who believed in the universalism of man and in a global community, the borders dictated by states would be an insult to the fundamental right to come and go. Marx believed that nationalist sentiment would lead to a false consciousness invented to hide the economic domination of capital.

Nationalism has been linked to economic models and also to political disputes. The First World War was driven by nationalist passions. With the end of the war, nationalism would once again win, this time as a political principle. "Wilson's Fourteen Points" were a series of peace measures recommended by American President Woodrow Wilson to establish the principle of self-determination. This principle was used, initially, to solve the issue of new political borders of peoples liberated from old dynastic empires by the war (Russian, Turkish and Austro-Hungarian empires). "One people, one nation" was the *motto* of the time. It would not be easy. Establishing the new nations was immensely difficult. How to suddenly accommodate people in pieces of land and states that did not exist before? And what about ethnic groups who secularly lived side by side, would they have to be separated all of a sudden? This kind of contradiction manifests itself to this day, such as during the Balkan Wars of the 1990s.

In Africa, the adoption of the self-determination of peoples was a disaster, done on the borders between old colonial empires. A political model and exogenous borders were both imposed on heterogenous populations. Many nations have been drawn with ruler and square as space delimiters. In other countries, tribes from the same ethnic group were divided between different countries and enemies were placed together inside new borders. The result was chronic political instability, civil wars, and genocide. With fragile states, enormous social unrest and environmental degradation quickly emerged.

Soon after the independence of India, something that struck me was Mahatma Gandhi's reaction to a question from a journalist: "do you want India to have the same standard of living as England?" To which Gandhi answered "It took Britain half the resources of this planet to achieve its prosperity. How many planets will India require for development?"

Gandhi was an environmentalist, perhaps unknowingly, perhaps knowingly, but the fact is that his words have announced a challenge we are now able to see clearly: the coexistence of economic growth and environmental balance. Gandhi foresaw the problems of sustainability of our ways of living, consuming and producing.

The European Union was a promise to create a new political organization and has been suffering with economic crises and differences among its countries in relation to their foreign policies. The EU is not a sovereign entity acting jointly on international affairs. The European Union meant the construction of a new political project, still experimental and specific to that continent, a new way to organize states and societies. However, the process of political union is incomplete – European states often do not agree and do not share a common foreign policy, which was evidenced in their different positions regarding their support of the American invasion in Iraq in 2003.

Economically, there has been strong integration since the 1950s, but the monetary union has been going through systemic crises, because there is a distinct asymmetry between core countries such as France and Germany and peripheral ones such as Greece, Ireland, and Portugal, the promises of fiscal balance and control of public deficit of those three countries not having been fulfilled. Still under the effects of the 2008 crisis, the euro has not become a viable alternative to the dollar. Monetary union is fragile without fiscal union.

Even with all those crises, it is the model of nation states dominating the global political landscape. Why this trend toward national value above broader and more comprehensive values? Perhaps because the concept of nation has some concreteness in our collective imagination, while the concept of humanity is vague.

A great motivator for me while writing this book has been reading and rereading. Émile Durkheim, like Isaiah Berlin, deserved to be back on my desk. Durkheim said societies suffer through the process of modernization, at different times and in different ways, but that this suffering is inevitable. The act of "modernizing" means destroying traditional social categories, replacing them with state centralization and bureaucratic and administrative rationality. The change was deep and traumatic for those who underwent that process. Leaving a peasant village for a city; shifting one's loyalty from a local leader to an abstract state which treats one impersonally; breaking provincial rules to abide by national laws: those are cognitive and emotional processes bearing enormous cost.

Loyalty in relation to new forms of organization was not instilled in people simply by convincing them, rationally and philosophically, that it was good for them. Nationalism was the solution to fill this void, because it is the emotional bond between state and individual. The father-state, the protector-state, the saviour-state, the provider-state.

There is no political doctrine which has not explored to the fullest the sense of nation. Historic opponents did it in the past and still do it in the present. The political world has taken nationalism to extremes, which Isaiah Berlin called expanded nationalism: the process in which the life of an individual can only be meaningful when tied to the history of his nation. The group starts to shape the individual; the state becomes an organism whose proper functioning depends on the behaviour of its citizens. Human essence itself would become linked to national unity, rather than to a family or clan. According to Berlin, individuals are like leaves from a tree: they may fall and be replaced, but the tree remains standing. Images like this came to be used by the propagandists of nationalism, especially in the last decades of the nineteenth century – and it worked very well.

Of all human passions, the most dramatic and anachronistic is perhaps the willingness to kill and die for a cause. We saw crowds willing to die for their homelands, even if many did not understand exactly what was behind the conflicts between nations. In the Napoleonic wars, soldiers were elevated to the status of demigods. Nationalism was ingrained in humanity in a striking way. The Enlightenment philosophers were counting on the transience of nationalism toward a global human community. What would they say?

THE CROOKED TIMBER

"Aus so krummem Holze, als woraus der Mensch gemacht ist, kann nichts ganz Gerades gezimmert werden" (Out of timber so crooked as that from which man is made nothing entirely straight can be built)."[4] Isaiah

Berlin was the great popularizer of Kant's phrase, which became the title of the book I referred to earlier – *The Crooked Timber of Humanity*. Yes, that is right: this thought has been with me for years. Nationalism is the crooked timber of humanity.

We need to rethink nationalism as a political principle if we are to seriously address the environmental issue, which concerns all of us.

We shifted almost all our loyalty to this abstract entity we call nation. Throughout this chapter, I would like to show some dilemmas and current impasses from this sort of human imagination. Disputes at sporting events between nations are a shining example of these emotional manifestations. How bloody nationalist passions are...

Nationalism is so familiar that we are not surprised by disputes between nation states, which almost always consider only domestic issues. This is the main reason for the difficulties in achieving a political agreement that benefits humanity as a whole, for example, at the climate conferences organized by the United Nations. What I have seen at the various COPs (Conferences of the Parties on Climate Change) are national particularities prevailing over the larger interest to protect the natural systems on which we all depend. I will return to that topic in Chapter 7.

The nations will have to relativize their own immediate interests to seek articulations and be able to reach agreements. But how can we create conditions for various national interests to cooperate, in the best interests connected to supporting life on our common home, the Earth?

DEMOCRACY AND POWER

The modern state is the consequence of a process of political and economic development, and one of its goals was to ensure continuity in terms of trade. Political philosophers of the seventeenth century have dealt with this relationship. To Locke, one of the functions of the state is to guarantee the right to ownership, while to Hobbes its primary function is to prevent the state of *bellum omnia omnes* (war of all against all). Despite their differences, both thinkers bring forth the idea of a contract, which would bring order to social and political relationships.

The Industrial Revolution accelerated the interdependence between politics and economics. With demographic and urban growth, the economic factor has become the main force driving the structure of the state. Protecting domestic markets became a mission of the body politic of the nationalist project.

Technological evolution and huge economic growth in the twentieth century broke borders, and forced men to rethink their political values. But what is an international society? What would the relationships between so dissimilar countries be like?

In 1990, in the article "State, Peace and Power: an exercise on democracy," for the book *For a Secure Tomorrow*,[5] published by Peace Initiative in Mumbai, I discussed about two political theories of international relations: Hobbes' and Kant's models.

For Hobbes, in international relations the only law is the survival of the fittest. International politics is power politics, it is self-justifiable. In contrast, domestic politics needs to be internally enforced to ensure social order.

Kant's viewpoint is universalist, his main idea being that of peace guaranteed by a benevolent state, comprised of legitimate representatives of the will of the people. Today, rethinking both theoreticians, I evaluate that we are living in both worlds. In foreign relations, countries follow the Hobbesian model, in internal relations, the Kantian model.

But in order to maintain internal peace in the Kantian sense, freedom and co-responsibility of citizens are required. In a democracy, the goals of society must be pursued harmoniously and in a participatory manner. Decisions taken by a majority have to guide the political process, whilst always respecting the rights of minorities.

While the West faces structural crises, China, an undemocratic country, is taking charge of the scene.

Not long ago, we used to imagine two trends. Chinese economic growth would raise democratic opposition, which would destabilize its one-party system, or the governance model of the Communist Party would not be efficient enough to continue leading the process of economic growth. None of these hypotheses have become reality. We do not know until when this model will remain viable in the pursuit of power by the Chinese. But what is power?

Power is the ability to control the behaviour of others. It should be used as a tool to accelerate social processes toward democracy and quality of life, not to restrain them.

Democracy means not only a system of civil and political liberties. To achieve the state of peace envisioned by Kant in domestic and international relations, the key is that people need to feel they have a purpose in life and are able to participate in the construction of their own future. There must be a feeling of belonging to something bigger; life should not be only about working-consuming-resting.

Individuals from different nationalities and cultural strata, especially the younger generations, are increasingly mobilized for causes rather that political parties. The environmental cause is the best example. The left-right polarization, so strong in my generation, makes less and less sense. There are only choices between models, one either defends a more liberal scheme, with a smaller state (Anglo-American-inspired), or a stronger state, with more social protection (continental European model).

The proposal for sustainable development is political in its essence, because it is related to a new way of thinking and acting.

FOR A NEW GOVERNANCE

There is a crisis of governance. This word was gradually incorporated into daily practices and political discourses. In the same year of the Earth Summit, the World Bank, in its 1992 document "Governance and Development," defined governance as *"the manner in which power is exercised in the management of a country's economic and social resources for development."*[6] The expression appeared due to the concern in shifting the focus from the economic actions of the state to a broader perspective which included both public management policies and social policies. Governance is not the same as government – it relates to methods and processes which produce effective results.

The model of governance must be changed and its democratic character expanded. State interference in the economy and social organization can be positive or negative. And here we return to the governments' decisions, always behind closed doors. These decisions can be positive when the state is willing to interfere in economic activities that harm the environment. When the state fails to fulfil this role, having no interest or strength, it runs the risk of contracting debts and environmental impacts, and everyone will have to pay the bill. Competition between countries and companies on a global level hampers the rational and sustainable exploitation of natural resources.

Environmental degradation, which until recently was considered at the regional level, became global. The gases released into the atmosphere and pollutants dumped into the sea do not respect political borders or international agreements. Environmental services such as carbon capture and supply of moisture to the atmosphere, carried out by tropical forests, are not only important for Brazil, Indonesia, or the Congo, but to the whole world. In July 1997, in an article for the newspaper *O Globo*, I drew attention to:

while man penetrates the fantastic adventure to unravel the mysteries of Mars, 497,000,000 kilometres away, deforestation and illegal logging make the Amazon Forest an increasingly bleak place. Five million truckloads of wood are taken each year from the region, which, together with the depletion of Asian and African forests, in less than three decades will have already become the main centre of tropical timber. Without information there is no accountability. Part of the problem is due to disinformation of the public – used for generations to associate the Amazon to the "abundance" and "inexhaustibility" of natural resources – only now beginning to be bombarded with news about its destruction.[7]

In this same article, I remembered that, in the wake of reports on this increasing destruction, according to a SAE (Brazilian Intelligence Agency) document, 80% of Amazon timber extracted from the Amazon had predatory and illegal origin. At that time, the government rushed to announce that models for environmental management were being studied. On the eve of the third millennium, it was unacceptable that in the vast Amazon region a legal jungle continued to reign. Does this sound up-to-date?

THE ULTIMATE FRUIT OF THE ENLIGHTENMENT

In the lecture "Global Governance in the 21st Century" at the London School of Economics in 2004, Fernando Henrique Cardoso observed that information technology has empowered individuals and communities, improving the conditions of citizenship, extending the public space, and we could even talk about democratic synergy between democratic nations, driven by technology.

The legitimacy of governments is no longer dependent on ideology or causes, but on the capacity to offer citizens what they expect: "Ideas and expectations now travel fast and can flourish wherever sensitive minds are available to pursue them. [...] Consumption styles became global, regardless of cultural or national differences. [...] governments are asked to deliver services at a pace where they were not used to." affirmed Fernando Henrique.[8]

Today the *motto* is not *what* to do, but *how.*

The following year, in 2005, with "The Need for Global Democratic Governance: The Perspective from Latin America," Fernando Henrique argued that "democratic governance, be it local or global, will always be a gradual, painstaking construction of the intellect, of experience, of rationality applied to a concrete and complex reality. Democratic governance may be the ultimate fruit of the Enlightenment."[9]

A deep desire for renewal is behind the movement toward democracy. People experience a growing sense of political freedom and want to have more voice, at the same time as distrusting traditional politicians.

In Brazil, political parties and Congress are among the institutions that people distrust the most. In other democracies, the picture is not very different. Traditional forms of political representation are today at stake.

Individuals today have many interests extending beyond the identity of a nation. Personal preferences count more and more in the formation of new identities. For this reason, nations do not carry the importance they have in the past. This could be the beginning of the end of the "crooked timber" mentioned by Isaiah Berlin.

Other forms of political participation have been gaining strength, such as direct communication with authorities, participation in non-governmental organizations, virtual forums for discussion, instant information on the internet, expansion of consumer rights. The platforms are many and the number is increasing.

This new reality also applies to the environment. The challenge of democracies today is to know how to adapt to this new rapidly-changing reality.

During the Cardoso government, in 1997, we environmentalists had great expectations for the Kyoto conference, which we assumed to be an institutional landmark. Kyoto may not have been the success we imagined, but I remember well the enthusiasm we felt with that meeting. In September, a few months before the meeting, I wrote the president:

I refer to the position of Brazil at the Conference of the Parties in Kyoto on Global Warming in December this year. The Conference in our view as scientists and environmentalists will be as important for the next fifty years as the Bretton Woods Conference was for the last fifty years. Contrary to what it may seem, Kyoto will not be one more diplomatic meeting of environmentalists, but the beginning of the shaping of a new economic-financial system the planet should take on during the next fifty years. This subject is causing, in developed countries, very strong confrontations between "lobbies" which represent the interests of the oil industry, the energy matrix, and the automotive industry against the positions of the governments of those countries, based on increasingly demonstrable scientific evidence.[10]

I got a response from Fernando Henrique about the expectations for Kyoto:

"There are great expectations regarding the world meeting and, as could be expected, this has been a constant theme in my recent conversations with foreign leaders, including President Bill Clinton, to whom I had the opportunity to mention the specificities and clean character of the Brazilian energy matrix. The American president too gives priority attention to the issues that will be discussed in Kyoto, seeming to be aware of the need to facilitate the transfer to countries like Brazil of environmentally healthy technologies.

The discussion in Kyoto, which gained urgency in the light of mounting evidence of climate change, will certainly be a strong force in shaping the economic development models from the turn of this century. I assure you that Brazil is ready to play the role it deserves in the process of decision making."[11]

WHAT DOES BEING INTERNATIONAL MEAN?

International bodies exist only thanks to the approval of national states. The very expression "international" already leads us to the reality of living under the aegis of a system that is not truly global, but between nations.

The most global institution, the UN (United Nations), has no real power or sovereignty. The UN follows what its member states determine and is therefore a reflection of the expectations of these countries. The UN military forces are trained and financed by the countries that comprise it. Without their own army and without their own resources, with no power to collect taxes, the United Nations are more of an inter-country forum than a supranational power. Their biggest influence is in the area of *soft power* (power from prestige and influence, not from weapons), but still with many limitations. The UN can, at best, create agendas for its member countries.

So, the UN's authority is in the hands of states. And here, again we live with a paradox. There have been many declarations of rights during the last sixty years. However, many countries have disrespected those rights anyway, without facing any sanctions except for entries in reports, such as those from Amnesty International.

If human rights are still largely written, but not effective, what can we expect for the rights of the environment? We have a long way to go, both in politics and in ethics.

The UN institutions have been multiplying over the last decades. The environmental agenda has relative importance in the framework of international organizations. If we observe those institutions under the criteria of the triple bottom line of sustainability, we will see that institutional structures still use a logic from the past. An entanglement of acronyms form the mattress of bureaucracy, without considering integration between the social, the economic and the environmental. The United Nations react and warn about the environmental crisis, but still stuck within a governance model that does not meet the demands of our time.

On the initiative of the UN environmental agencies, WMO (World Meteorological Organization) and UNEP (United Nations Environmental Programme), the IPCC (Intergovernmental Panel on Climate Change) was created in 1988. This group is formed by a body of researchers who review and consider the importance of scientific and socioeconomic information for understanding climate change. The information is consolidated into a summary for decision makers, subject to the approval of governments. I estimate that this is a merit of the IPCC, to connect governments to techno-scientific results. This is usually the great critique of the ones who are sceptic in relation to the IPCC, but therein lies its virtue, because otherwise it would be just another scientific report among thousands that are published each year.

The first IPCC report, published in 1990, indicated that climate change should be part of a political platform between countries to avoid major consequences. This alert played a decisive role in the creation of subsequent treaties on climate. In 2007, the IPCC published its fourth Assessment Report, which is considered the most detailed summary on the issue of climate change. In it is declared that the climate system is unequivocally warmer, and that much of this increase is the result of an accumulation of greenhouse gases with anthropogenic origin. In 2007 the IPCC, led by Rajendra Pachauri of India, has become the most important reference about climate issues and shared the Nobel Peace Prize with Al Gore for his documentary *An Inconvenient Truth*.

Due to the precautionary principle, the IPCC forecasts in its 2007 report were relatively optimistic. In recent years we have seen that the intensity and speed of ongoing climate change are greater than predicted by the report.

In the field of multilateral negotiation for the environment, the United Nations organized the Conventions, which depend entirely on the cooperation of the countries to take place. As noted, the Kyoto meetings meant the recognition by nations that a problem really existed. However, in the country which was then the largest polluter, the United States, the Kyoto Protocol was blocked by strong economic lobbies, especially those linked to energy and transport, urging Congress not to ratify the agreement.

Decisions between countries gathered in COPs can only be taken by consensus. So it's not difficult to imagine that this is a slow and inefficient process.

And here we return to nationalism. When they go to the COPs, delegates work with their domestic audiences in mind, their interests at home. Insignificant intricacies in the final agreements are discussed as if a small passage of the document could decide the entire climate problem. The COPs reveal the difficulties existing in democracies and in the model of traditional political representation. The United States are the worst case due to the difficulties in obtaining congressional support in relation to multilateral decisions. The COPs end up functioning more like a political stage than as a forum for environmental agreements.

There is no possible way to a solution if most major polluters do not take responsibility.

A NEW INSTITUTIONAL ARRANGEMENT

We must seek a new institutional arrangement in order to balance the sovereignty of the states with global needs. The best way, I think, is to avoid repeating the nationalist model, explained in this chapter. International agreements, for example, are only really effective when they become internal laws in the countries that signed them.

There is a broad international institutional order, yet it lacks the strength to act on rules and regulations essential in halting climate change and preserving natural resources.

Civilization at times has had to rise around great empires and their areas of influence. For good and for evil, empires indicated the direction that the world, or the region under their influence, would take. In recent history, we can remember the British Empire and, during the Cold War, the ideological empires of the United States and the Soviet Union, which fought for political hegemony and developmental models. As there is no longer a single dominant empire in the world, the relevant thing now is to coordinate efforts so that the threats of environmental devastation are effectively considered by the global community, regardless of their ideologies.

Our institutions must change, emphasizing the democratic governance model, bottom-to-top, focusing on the participation of peoples and coordinated efforts to promote the changes we so badly need.

It is the natural capital that will ensure life for future generations. Climate change creates an enormous challenge to the established political model. But why don't politicians see the obvious?

POSSIBLE FUTURES

We are billions of individuals who demand efficiency and productivity in order to obtain resources to ensure our survival in the short term, while the production of wealth, the natural capital, occurs in the long term.

I would like to mention another one of my old teachers, a person who greatly influenced me during my student days in Paris, the philosopher and political scientist Bertrand Jouvenel. He adapted the expression *Futuribles*, by theologian Louis de Molina, a terminology that came from the contraction of the expression *Futures Possibles*. Jouvenel believed that there are many possibilities for the future and was against any kind of determinism.

I remember hearing him in Paris, during a conference in which he affirmed that society, in its dynamics, adapts quickly to new technologies changing daily habits. An example of this quick adaptability today, I might add, is the widely assimilated use of the Internet and mobile phones. But when it comes to ideological changes and political structures, the chronology of these changes takes at least a period of two generations to become part of the public consciousness.

The environmental crisis does not discriminate between countries. The liability of state actions begins to extend beyond borders. Nations more vulnerable to climate change, like those in Micronesia (an insular region in the Pacific Ocean), are already seeking to prevent the development of dirty energy projects that may cause more damage to the planet. An unprecedented event occurred in May 2011, when the Federated States of Micronesia requested an assessment of environmental impacts caused by the expansion of a coal plant in the Czech Republic. Projects involving carbon emissions could be challenged in the courts. Nations affected by climate change are beginning to use international law against the major carbon emitters.

One of the most perverse effects of climate change is the rise in sea levels which, in those countries, has compromised agriculture and the supply of drinking water. Internal migrations are already a reality, and programmes to displace population within twenty years are already under discussion. The 21st century refugees will be the victims of global warming.

Countries with great economic needs are in general those with fragile political structures, with little effective governance practices. Countries with low CO_2 emissions will suffer the most severe consequences of climate change.

STATE: BAD WITH IT, WORSE WITHOUT IT?

There is still no substitute for the state. Consequently, it is possible to observe today the phenomenon of failed states, countries in which the formally established government does not actually control the territory or does not have the strength to impose itself on factions. These are torn countries, economies in regression and anomie, and a complete failure of basic rules between individuals.

Currently, perhaps the worst examples of failed states and its consequences are in Somalia and Sudan. Somalia has lacked a functioning government for decades. It is politically divided into two regions: "official" Somalia and the region of Somaliland, in the North. The irony is that the latter, which wants to have its independence recognized, has a relatively stable government able to guarantee peace, while in the South, the remaining part of Somalia which is recognized by the UN, is a disaster in political, economic, social, and environmental terms.

Under civil war for over twenty years, Somalia is a portrait of a failed political model, an archaic model of social organization and an economy which depends on degraded land and low productivity.

The groups fighting each other in this civil war are from old tribal clans and, even if they are mostly Muslims, old hatreds stand out due to lack of resources. The traditional tribal wars fought with bows, arrows and swords are now fought with Kalashnikov machine guns in the hands of teenagers. The groups fight for water, pastures and cattle, and for control of the main city, Mogadishu. The officially recognized central government remains entrenched in Villa Somalia and controls a few parts of the capital.

It is a humanitarian catastrophe and a symptom of political and economic bankruptcy. Between 1992 and 1995, two military interventions by the United Nations took place, under the leadership of U.S. forces, which remained for a short period of time in the country and then withdrew because of the powerlessness of their actions.

The nature of political power abhors a vacuum. Where there is vacant space, soon new forces will be established to conquer it.

The lack of economic perspectives and widespread conflict have led to the reappearance of piracy on the high seas. It seems unbelievable, but there are pirates in the twenty-first century, who, with motor-boats, machine guns, and GPS, attack large freighters in a key shipping route for global commerce, connecting Europe to Asia. The pirates are the paroxysm of a failed political model. Their leaders are heads of self-governing militia clans from a territory near the Somali coast, attracting young people who are unemployed and without prospects. The attacks and hijackings of commercial ships have generated income, which is largely reinvested in more weapons, fuelling the civil war machine. On July 20, 2011, FAO officially declared a state of famine in two regions in the south of Somalia. A combination of institutional failure, draught and high food prices leaves millions of Somalis in dire need. The extreme example of Somalia may serve as an alert about the social, political, economic and environmental conditions that drive human beings back to barbarity. Twenty years of anomie in a country were enough to produce wars, terrorism, and piracy. This seems to be a history that is far from us, but there is no ontological difference between us and the Somalis. What happens there is the sum of undesirable products of contemporary history, political models that are not adjusted to local reality, population growth without higher productivity, and scarcity of water and cultivable land.

In the distant past, Somalia was one of the jewels of Africa, a bridge between the Arab world and pagan Africa. During the Middle Ages, Mogadishu controlled gold routes that supplied trade between East and West. Its canals and dams were examples of hydraulic engineering and technological inventiveness. The current disaster can be seen as an experiment about what humanity is capable of in extreme situations of survival. It serves as an alert to the search for a political model that considers economic and social needs.

The democratic state, with all its shortcomings, is still the best starting point for thinking about the improvement of our political model. We are increasingly in need of globally controlled policies, without nationalistic hysteria, without chauvinism, without interference from private interests, thinking and planning our future together, combining the urgencies of the present with the needs of future generations.

Politics can no longer be the axiom of the Prince of Falconeri, Lampedusa's *The Leopard*, (*Il Gattopardo*):[12] "If we want things to stay as they are, things will have to change." This remark became well known because it applies to the politics of many countries. They make changes that ultimately do not change anything, reforms are made to maintain the *status quo*. While there is no broadening of thought towards changes of paradigm, everything will change in order to stay as is. Until when, princes of Falconeri?

THE GREAT CHINESE LABORATORY

In our time there is no one who does not recall that day in November 1989 when we witnessed the fall of the Berlin Wall. The images were right there in front of us. We exchanged phone calls trying to understand that unimaginable fact. Not long after, a current of political thought claimed the end of history. The argument was that the collapse of the Soviet-Communist bloc would bring the world to a solution in terms of political-economic model. Liberal democracy would gradually expand to the entire planet, bringing promises of prosperity and freedom to all citizens. How naïve.

Who could have predicted the immense role that China would have in the world today? Under its own model, it combined centralism and Confucianism with the speed of capitalism. The largest laboratory on the planet today is called China. It is unique in many ways. Now, as I write, there are 1.3 billion Chinese. Slightly more than half still live in rural areas, although in the next decade, this should be reversed. China has a huge middle class in constant growth, the equivalent of two Brazils, which will require more and more food and jobs. The Chinese population has shown increasing rates of consumption.

China also experiences the largest process of urbanization and industrialization the planet has ever witnessed. We are already experiencing, and will experience more and more, the consequences of the Chinese model. If they consume food with the same voracity as Americans do, they will need two-thirds of the current production of grain just to feed their cattle, whose size would be 80% of the current world herd. In regard to energy, there would hardly be enough: if China consumed oil and coal at the same rate as the United States, the world would have to produce twice as much of these resources in relation to current production, equally doubling the emissions of carbon dioxide. If there were the same number of cars per inhabitant in China as in the United States, the area used for roads, highways and parking lots would be equivalent to the current area used for rice plantations.[13]

The Chinese rely on a dual strategy. Economic growth, based mainly on coal, makes China the biggest emitter of greenhouse gases on the planet. But at the same time, it is the largest investor in the development and production of clean energy sources. China's emissions are what moves good part of the world economy.

The pace of economic growth in Latin America during the last decade (2000/2010) happened mainly due to the Chinese demand for minerals and other commodities. The invasion of cheap Chinese goods increased consumption in many Western countries, including the United States. So, what is happening in China is not only their problem. Several economies benefit from the *Chinese way of life*.

China has adapted to the model of globalization with an amazing speed. In ten years, from 1996 to 2005, the Chinese more than doubled the emissions of carbon dioxide from burning fossil fuels. During the Kyoto negotiations in 1997, no one considered China would become the largest emitter of greenhouse gases.

Its influence on the world expands as much as its emissions. Only in 2010, the Chinese have announced investments of over 11 billion U.S. dollars in Brazil.[14] Steel production, oil exploration, electricity distribution, mining, transportation, these are some of the sectors in which they want to invest.

Companies and state enterprises are engaging in the race to supply China with the raw materials they need. The editorial of the newspaper *O Estado de São Paulo*, of August 2010, already alerted to the fact that the Pallas International Corporation, formed by Chinese private and state investors, announced plans to buy between 200,000 and 250,000 hectares in western Bahia and also in Maranhão, Piauí and Tocantins, a region known as Mapito. The largest state-owned company in the food sector, China National Agricultural Development, already operates in forty countries, and ten thousand of their eighty thousand employees work abroad.

The Chinese have also entered Argentina, Peru and Africa, bought land in Tanzania, Guinea and Benin, with the same goal of ensuring food for its population.

In practice, China has not yet reached the status of a free-market economy. State and private interests are articulated in an unclear kind of game. There are more guesses than certainties, since the Chinese State is not transparent about its actions.

In the Chinese perspective, history between the seventeenth and twentieth centuries was a parenthesis of their global leadership. Throughout the Middle Ages, they were the most technologically advanced, dominating the seas and having control over trade routes. Its political model of centralized empire already presupposed meritocracy and hired employees through public tenders. At the turn of the eighteenth to the nineteenth century the emperor declared that the Chinese had no need to consume the manufactured goods Britain wanted to sell them. It took two opium wars for the English and, with them, other Western nations, to dominate the markets, territory, and even the government of China. After these conflicts, a national question began to take form: how to reach and surpass the West?

The common idea in China is that a unitary power is necessary to ensure order. Yesterday, the emperor. Today, the party. Certainly the Chinese walk "on two legs," as Mao Zedong used to say in other times, but now we can say: one leg is Eastern and the other Western.

WE CAN NOT REGRESS

Total war, a concept used by Eric Hobsbawm to describe the world wars of the last century, is the worst possible scenario for a globalised and very well armed planet. We know the dangers that a future shortage of water, land, and food could bring to political stability. According to Hobsbawm, total war means humanitarian regression: "Certainly both the totality of the war efforts and the determination on both sides to wage war without limit and at whatever cost, made its mark. Without it, the growing brutality and inhumanity of the twentieth century is difficult to explain. About this rising curve of barbarism after 1914 there is, unfortunately, no serious doubt."[15]

Genocide (a neologism invented during World War II by the Jewish refugee Raphael Lemkin) and the masses of refugees can be considered modern inventions, just like airplanes and broadcasting.

Total war was also harsh against nature: forests and agricultural lands were turned into bombed deserts. It makes no sense to repeat all that. Negotiations between nations are inevitable. When we talk about globalization we are not talking about a world government, but rather about an intertwining of national destinies.

CHALLENGE-AND-RESPONSE: ON A MOTOR-BOAT WITH TOYNBEE

To conclude this discussion about the political model, I would like to comment on the consideration of Arnold Toynbee, a thinker I greatly admire whom I met in the late 1950s on a boat trip, in the company of Roberto Campos.

Toynbee wrote great part of his work in the period between 1920 and 1950. Therefore, both world wars marked his intellectual journey. There is a passage in his most important book, *A Study of History*, which, in a way, sums up his political theory:

"Every society is faced, in the course of its life, with a succession of problems, which every individual has to solve by himself the best way he can. (...) As difficult times succeed each other, some members of society (that is, some states) cannot adjust, and get lost on the way; (...) others grow in wisdom and greatness and, by taking their own directions, discover new directions for the general process of the society in which they belong."[16]

We can extend Toynbee's observation to our time. We experience a succession of problems, the scientific community points toward several solutions for environmental degradation, but still there is not a common path. And, in this exercise in historical parallels between past and present, the main difference now is the shortening of distances through technological advances.

What is the relationship between the past and the current issue of Earth's sustainability? There is a close relationship between access to natural resources and political and ideological disputes. Conflicts became gradually more destructive. The Second World War marked the end of European hegemony over the world, meaning the failure of the civilization projects Europeans were aiming at.

The motivating process in the success or failure of societies appears in Toynbee's concept of Challenge-and-Response, from *A Study of History*. The social elites of that period did not manage to come up with creative solutions to the problems they were facing and sank into militarism and ideological archaisms. Toynbee affirms that:

"One of the main impulses (of regression) is the virus of nationalism, which we can see acting in the contemporary world. A community that succumbs to this serious spiritual malady is bound to create a parochial national culture, which can be declared free from foreign interference."[17]

The environmental crisis is a consequence of political models which no longer serve us.

Toynbee makes us reflect: "The greatest punishment for those who are not interested in politics, is they will be governed by those who are interested."

5

My experience as Mayor of Rio de Janeiro

"You take delight not in a city's seven or seventy wonders, but in the answer it gives to a question of yours. Or the question it asks you, forcing you to answer, like Thebes through the mouth of the Sphinx." Italo Calvino, in *Invisible Cities*.[1]

"You have to take office in Rio." The invitation came via telephone from the then Governor Chagas Freitas and from the Generals João Figueiredo and Golbery de Couto e Silva, Chief of Staff and organizer of the political opening. I felt that if I accepted, I ran serious risk of being held hostage to an exchange of political positions for political favours. The invitation challenged me, but I would only accept to take office if I could form my own team. We hung up.

The next day, in the early hours of the morning, the governor informed me that I would be allowed to form my own team. The first battle being won, I still had another one ahead, a giant and much more complex one: I wanted to dissolve the ill-fated merger of the states of Guanabara and Rio de Janeiro, which had occurred four years earlier.

The municipality of Rio de Janeiro was undergoing a major financial crisis as a result of the merger. President Figueiredo told me not to worry; I should talk to Mario Henrique Simonsen, because he would release funds for the city. Figueiredo, knowing about my friendship with Mario Henrique, then the Planning Minister, told him to call me. "Israel, come here to talk with me because the matter is complicated. You will have to accept." I showed him that the city of Rio was broke. Figueiredo then said to Mario Henrique: "Solve this financial problem so Israel accepts once and for all." Mario Henrique managed to get a loan of 300 million U.S. dollars, to pay off debts. At that meeting, in Brasília, I returned to the subject of the dissolution. I reinforced again that it was my intention.

I took office in the city of Rio de Janeiro in March 1979, largely because of the influence of Mario Henrique, a friend since my Engineering School days, when we formed a group in frequent meetings around Dr. Eugênio Gudin and Dr. Octávio Gouvêa de Bulhões. Five months after I became mayor, Mario Henrique left the Ministry, due to conflicts with Delfim Netto.

Figueiredo continued Geisel's policy of political openness. Among advances and setbacks, Brazil followed the path back to democracy, and 1979 was the year of the amnesty and also of the second oil shock. Increasing international interest rates, foreign debt at floating rates, and fuel shortages were the ingredients of an economic crisis which lasted through the 1980s, dubbed the Lost Decade. It was in this context that I played the role of Mayor of Rio de Janeiro.

On inauguration day, the previous mayor, Marcos Tamoyo, left a letter on my desk, informing me that a teachers' strike was already scheduled for the week after my inauguration. Lack of money, overspending, frequent strikes... that gloomy financial horizon made it difficult to solve problems of health, education, and the chaotic growth of the city. The solution was to continue demanding the aid due from the federal government. I announced the deficit for 1980 and went to Brasília to talk with Delfim Netto.

With the merger, it was the duty of the Federation to support the formation of a municipal budget until the city's fate could be consummated and it could create its own resources. Rio de Janeiro, previously self-sufficient, after the merger had to share with the state the taxes it collected.

The financial problem stemmed largely from a political error. I exchanged grievances with Dr. Gudin. He pointed out that the merger had been decreed without any previous study regarding the impact on the structuring of the new state and the new municipality. Therefore, the evidence that the merger had been a mistake was overwhelming.

In this archeology of memories, a pleasant surprise was to rediscover old papers. One of them is an article by Dr. Gudin: "Israel Klabin at the helm of the rotten borough," having as epigraph the advice of D. Quixote to Sancho: "[In] that tempestuous sea, wherein you are going to be engulfed [...] offices and great employments are nothing else but a profound gulf of confusions." Dr. Gudin also brought up a thought on my uncle Horácio:

"The municipality is individually unfeasible and every year closes the budget with loans, overwhelmed with problems of education and health. (...) Your situation reminds me of an aphorism of your relative and dear friend, Horácio Lafer, referring to the Ministry of Finance: 'to have been is good, but being is hell.'"[2]

The city continued living on loans, and I thought fit to change its managerial mindset. The bureaucracy required for projects to be approved took more than six months. I asked for help from Hélio Beltrão, then Minister of Debureaucratization,[3] and signed a decree that ended the need for notarised documents in public administration institutions in the city. Until proven otherwise, every citizen was innocent. The individual had priority over matters.

I had invited Carlos Alberto Direito to be chief of staff, and he helped me very much with all the legal and institutional situations; I also formed a council of notables, which I called *Câmaras Técnicas* [Technical Chambers]. The COMUDES (City Council for Economic and Social Development) met once a month with me and my cabinet and was formed by Octavio Gouvêa de Bulhões, Glycon de Paiva, Hermelino Herbster Gusmão, Gilberto Paixão, Arlindo Lopes Corrêa and Luis Fernando da Silva Pinto. Later, Olavo Setúbal, former mayor of São Paulo, would join the Council.

The Technical Chambers were formed by a group of independent people who thought about directions for the city, away from the tangle of bureaucracy. The Chambers were topical, one for each sector of the municipality. There was an Acoustic Chamber which determined parameters and decibel limits for different regions, there was the Sanitation Chamber, and so on.

Until then, the Department of Works was a great empire: it projected, built and licensed the buildings of the city. The creation of *Fundo do Rio* [Rio Fund] helped manage public resources for investments in favelas. In 1979, the city already had more than 300 favelas, with more than 1.7

million inhabitants. We were watching the city being disfigured. According to the Board of Parks and Gardens, of every 100 trees planted, only fifteen were still alive after one year. In September 1979, we began the PROMAM (Programme for Environmental Protection), which was unanimously approved by the Municipal Council and voted into law.[4] At the time of approval, I recalled that

we have taken the first step towards providing the municipality with a specific law in defence of the rational exploration of our natural and renewable resources, and of the environment.[5]

The PROMAM was an emergency programme designed to reverse the process of deterioration of the quality of life in Rio. It had a total of thirty-eight projects. Among them, the creation of Bosque da Barra, in an area of 613 thousand square metres, where 300,000 trees would be planted. The PROMAM would function as a political basis for the environmental policy of my administration, and could and was intended to continue in future administrations.

We wanted the PROMAM to be more educational than punitive, so it could guide a more humane society in the future. We predicted, for example, the reforestation of hillsides, incentives to plant trees, and even community activities such as fairs to encourage distribution of seedlings and the training of agricultural clubs in public schools. Community participation, one of the main points of the programme, was called "planting under secured protection," in which the basic idea was to stimulate the highest possible number of community groups to participate in the projects, either by planting trees, or by ensuring their protection. With the same goal, we would extend the action to industrial areas. Still in the PROMAM field of action, the Municipal Agriculture Office provided seedlings to be planted in schools. A decree made it mandatory to plant trees in built environments and industrial zones.

I gradually matured the idea that environmental conservation should be extended to the concept of heritage. There was a persistent dissatisfaction with the destruction of the memory of the city. I began to form a more comprehensive view of the environment, linked to the notion of protection of historical and cultural heritage.

A city's historical view can be compared to that of a botanist when he observes the growth rings of a tree or an archaeologist when he excavates and exposes strata of cities that existed at some site. The whole life of a tree is written in its growth rings, as well as there are layers of different eras inscribed in the urban organism. Both in cities and trees, each ring or stratum represents part of their histories.

The municipal law was too permissive, which was driving the city toward huge deforestation and urban chaos.

One of the Technical Chambers was aimed at proposing solutions for heritage preservation. This group was responsible for bringing and implementing an innovative project for what remained of the memory of Rio. It was heartbreaking to see the city degraded, day by day.

The scenario screamed aggressively about the lack of contemplative exploration of land use in the past. I was shocked with the destruction which was occurring and making the city uglier. Rio was becoming devastated by the accelerated urban growth and the great works – both public and real estate.

I then invited the architect and urban planner Augusto Ivan de Freitas Pinheiro, who had done postgraduate studies in the Netherlands, to coordinate the project we would ultimately call *Corredor Cultural* [Cultural Corridor], the preservation of architectural aggregations located in the immediate periphery of the most dense area of the city centre. Soon another group of notables was formed for the Cultural Corridor Technical Chamber: José Rubem Fonseca, Rachel Jardim, Nélida Piñon, Lélia Coelho Frota, Sérgio Cabral, Ítalo Campofiorito and Paulo Alberto Monteiro de Barros (the King Arthur of our Round Table).

In Brazil, all this was very new. Years later, Augusto Ivan, remembering those times, observed that the debates "sounded a bit like what Mário de Andrade advocated in the time of the creation of the federal heritage agency: that the atmosphere of cities should be put under governmental trust."[6]

Although there was an agency dedicated to the preservation of historical heritage, the SPHAN, later IPHAN, the idea was to preserve monuments from the colonial era. Thus, the agency remained on the sidelines of the problems of chaotic urbanization. The city no longer had the traditional instruments for the protection of patrimony, such as a governmental trust. Therefore, we decided to enforce the implementation of legislation on land use and also to create tax exemptions promoting preservation of patrimony.[7]

I remember details in making the Cultural Corridor project come to life. The initiative was bold and, from the get-go, went through a trial by fire – literally. Carlos Alberto Direito and I agreed that we would start everything in absolute secrecy: "Let's make a decree, freezing the whole area that will be determined by the commission." We made the decree, very carefully so no one knew about it. But something strange happened: the houses of Rua Primeiro de Março, in the city centre, had been given a high valuation because the former mayor had given permission for the construction of buildings. A resident who owned two houses in the street knew about it somehow – it is hard to keep a secret like that – and set fire to the houses. On the same day I decreed the freezing. I also forced him, by means of a very heavy fine, to restore the whole façade and to keep the same patterns, using the inside of the building as he pleased, but the external part was all untouched. There was a lot of indignation, but we won in the end. The Cultural Corridor was born in the Department of Planning because it had as a principle a character of city planning, more than only preservation of historical heritage.

The document defined the Cultural Corridor as "the place where a cultural function was established in a continuous manner in the central area of Rio de Janeiro, according to historical, architectonic, and recreational characteristics." In the same document it was added about the area: "Its greater or smaller vitality will depend on the 'natural' expansion of its centre and on the urban models adopted by the agency or agencies responsible for the soil legislation and planning of the city."[8]

The architectural aggregations to be preserved stretched along an axis that began in Lapa, crossing areas of the Passeio Público, Cinelândia, Carioca, Largo de São Francisco, SAARA, ending at Praça da República. The second axis started in Largo da Carioca, crossing Rua São José, continuing to Estação das Barcas at Praça XV, encompassing the entire area between the end of Avenida Antônio Carlos and Rua Primeiro de Março and the sea (Cais Pharoux). Praça Tiradentes was also included in this area.

The Corridor surpassed its historic or architectural value. It took into account the emotional heritage of the city.

We also created tax exemptions in order to promote heritage preservation. The Cultural Corridor project excited the entire team at city hall. The writer José Rubem Fonseca was a very old friend from the times we served together in the Army. He always avoided public appearances, but suddenly began to give interviews, so excited was he about the idea of the Cultural Corridor. He then took charge of Fundação Rio, which encouraged traders, through incentive grants, to restore façades, returning them to their original appearance. The group implemented ideas and helped create cultural programmes. I well remember concerts in churches in the city centre, concerts at noon and at lunch time. It worked very well. Those are things that do not fade away.

The aim of the Cultural Corridor was to revitalise the city, increasing the self-esteem of the people, who suffered with the serious consequences of the merger of the states of Guanabara and Rio de Janeiro. The law was only approved four years later, in 1984, but the mindset was already consolidated. And the dissolution? The factor which motivated me to accept taking office, as we know, has not been accomplished.

Rio de Janeiro was once again reinventing itself. I ended up frustrated at not being able to dissolve the merger. I demanded many times a decision from the federal power, but I was told that there was no interest whatsoever to end the merger. I left my position on June 3, 1980, a painful farewell, after one year and three months as head of the city. I would take over, soon after, a hard nut to crack: the presidency of Banerj. But that is a different story, perhaps to be told in another book.

6

Economic Model

Never has humanity been so rich, but never has it been so much at risk of losing the foundations of this wealth – nature. We are used to considering it an unlimited and freely available good, which is simply not true.

There are several vectors that encourage the maintenance of an outdated model: 1) The non-inclusion of the concept of natural capital in the current economic model; 2) The inefficiency of regulatory mechanisms for global financial capital flows; 3) Models of development detached from local social and environmental realities; 4) The immense force of military budgets in the global agenda; 5) The anachronism of the GDP as an indicator of economic activity; 6) The low effectiveness of the Kyoto Protocol; 7) The non-inclusion of the carbon cost in the prices of goods and services; and 8) The fictitious character of current convertible currencies, the so-called *monnaie de compte*.

These eight vectors will be discussed throughout this chapter. But first, I would like to go back to the origins of the economic model.

We are experiencing a crisis of purposes. For Adam Smith, founder of the modern science of economics in the eighteenth century, the private is only good insofar as it serves the general, and follows in that direction guided by an "invisible hand," which controls the market economies.

For Adam Smith, the economy was an integral part of philosophy. He himself considered his work *The Theory of Moral Sentiments* superior to his best-known study, *The Wealth of Nations.*

In *The Theory of Moral Sentiments*, Smith considers the feeling of sympathy as a major element of the exchange system, which should always allow transactions which benefit both sides, like a game where everybody wins. The sense of *duty* is the pillar of virtue of the state itself.

Friendliness should not be understood here in its colloquial sense, but in the philosophical view Smith inherited from the ancients: a mutual attraction between members of a society, a willingness to trust and cooperate with one's fellows in light of the common good. It is in this sentiment of a philosophical origin of liberalism we should turn to in order to transform and update the current economic model. We saw this in Japan, at the time of the 2011 catastrophe. In an extreme situation caused by earthquakes, tsunamis, and radiation leaks, the Japanese people taught us a lesson in civility and solidarity.

Relations between nations have changed, but there is still a huge gap between reality and the actions of states. Economic and environmental crises are two sides of the same issue (would they be two sides of the same coin?), and we cannot waste the great opportunity we have today to create new policies.

Memory has an ethical dimension. I learned from my old counsellors who have always guided my life. My father and mother were my first north, those who gave me my direction, methods, and roots. Then mentors, friends, family, friends of the family, friends I made along the road.

For at least twenty years, I have been repeating in lectures and seminars the following thought: climate change is not the cause, but the effect of an open system in which the economic model has overbearing weight. This system has caused the current crisis.

The solutions should always be based on the principle that climate change belongs to this open system, in which there is an intimate intertwining of energy matrix, politics, economics, and ethics. They are independent forces.

MOTIVATED BY AN EMOTIONAL IMPULSE

The first example in history of a global monetary order was the Bretton Woods Conference, in 1944, which defined the management of the international economy.

The victory of the allies seemed already certain, and the asymmetry of power between the represented nations was huge. Bretton Woods was a unique event which, to this day, permeates our economic relations.

This conference led to the architecture of an economic model which continues to this day, with only a few small changes. The communist countries, led by the USSR, participated in the meetings, but did not ratify the agreement. China abandoned Bretton Woods five years later, in 1949, with the Communist Revolution.

The Americans were the ones running the show. The United States were not bombed during the war and were at the peak of their economic, political and military power.

The U.S. GDP had doubled during the war, and they produced more than half of the world's energy. They had giant reserves of gold and led international commerce – they were the only country with nuclear power. Their dominance was absolute.

The goal at that time was to ensure the world would not regress to the economic turmoil of the 1930s. The rise of Nazi fascism, a by-product of the Treaties of Versailles, interrupted the experiment with freedom and humanism of the Weimar Republic. The economic effects of those Treaties led to Nazi fascism, and Bretton Woods sought, therefore, to be an anti-Treaty of Versailles, through the formulation of a new model that avoided the economic chaos of the 1930s.

Stabilizing a new order, in which the Americans would be at the same time main players and arbiters of the system, was their goal at Bretton Woods. This goal was reached and definitely reinforced American economic, military and democratic hegemony.

Brazil participated in this council. At the head of the great Brazilian delegation were two great masters, key figures in the Brazilian economy, already mentioned in this book: Eugênio Gudin and Octávio Gouveia de Bulhões. Roberto Campos, also present at the conference, would remember Bretton Woods with scathing humour, fifty years later:

"A sleepy village in the mountains of New Hampshire, where work would be the only possible distraction (...) although modestly, I was one of the protagonists of this formative period (...) somewhat wistfully, I was celebrated, in the fiftieth anniversary of the Conference, as one of the 'seven survivors,' something that sounds a bit like a movie title."[1] Where now will I be able to find that scathing humour and shy affection of Roberto?

Brazil contributed in Bretton Woods with an interesting proposal, which was defeated. Roberto reported: "instigated by Gudin, [Brazil] presented a seminal proposal, whose non-acceptance at the time made the international monetary system unbalanced and explains part of the bitterness currently surrounding the confrontations between industrialized countries and primary producers, in the North-South dialogue."

The Brazilian proposal defended the prices of commodities, basic products, increasing relations of exchange between countries and stabilizing the exchange rate of those which depended on the export of primary products.

The truth is that Brazil had no power to implement this type of argument. The developed countries pursued their own interests above the balance of international trade. This order stemming from the economic model is what guides many negotiations until this day.

The United States allowed, very democratically, for all delegations to speak and give their proposals, as long as the content of the final agreement was that approved by the Americans. At the beginning of the activities, there was already a previously agreed proposal between United States and Britain. The remaining countries had a role in helping with the technical work and giving an air of legitimacy to the agreement. The Americans dominated the world economy, but needed allies to have free access to the international markets.

I shall go back to Roberto in his "lantern on the stern" view: "Britain had already complied with the prospect of being for a long time a debtor nation. All assets, investments and reserves had been largely settled in the war effort."

The great star of the event, however, was not an American, but an Englishman, John Maynard Keynes. His unorthodox economic theory, preaching state intervention and public spending as antidotes to economic depression had helped the United States themselves overcome the crisis of 1929.

Keynes was no friend of *laissez-faire* style capitalism. In an article in *Yale Review,* in 1933, he wrote:

"The decadent international but individualistic capitalism, in the hands of which we found ourselves after the war, is not a success. It is not intelligent, it is not beautiful, it is not just, it is not virtuous – and it doesn't deliver the goods. In short, we dislike it, and we are beginning to despise it. But when we wonder what to put in its place, we are extremely perplexed."[2]

The economists, led by Keynes, wanted to avoid a currency war, the devaluations that occurred in the 1930s intending to prevent the concentration of reserves in the hands of creditor countries. The United States was the largest international creditor at the time. Keynes predicted the creation of something similar to a World Central Bank, with authority to issue a supranational currency – the Bancor. The monetary reserves of all countries would be converted to Bancor and deposited in a central agency, which would redistribute the liquidity to the international trading system. It was a proposal for the future that could have prevented many imbalances that afflict us today in the exchange rate system.

But the system that prevailed was that of the U.S. dollar as base currency. The country holding large reserves, credit and military power would be the only country able to issue the reference currency, backed by gold.

Bretton Woods also defined the postwar financial institutions: IBRD (later part of the World Bank), IMF, GATT (an embryo of the WTO). With their qualities and imperfections, those institutions shaped the global economic system.

NATURAL CAPITAL

How much do natural resources cost?

Herman Daly, a pioneering critic of the conventional model, sees the economy as "an open subsystem of earth's ecosystem, which is finite, non-growing."[3] Daly argues that, to the extent that this subsystem (the economy) grows, it incorporates an increasing proportion of the ecosystem. For him, the term *sustainable growth*, when applied to economics, is a paradox in terms – contradictory as prose and non-evocative as poetry.

Given the relative depletion of natural capital in developed countries, what we see is the import of capital in developing countries: wood, minerals and food. The most damaging form of degradation is still the indiscriminate exploitation of natural capital, which generates unequal patterns of consumption, and unsustainable transfer of wealth.

There are two aspects in the relationship between environment and poverty: a) poverty causing degradation through deforestation, predatory use of land, and rapid population growth exacerbating the pressure on the natural resources of national economies; b) degradation causing poverty through desertification, drought, and exhaustion of soil.

There are as many differences between the concepts of development and growth as there are authors. Whatever the theoretical framework, there is no end to environmental degradation without putting an end to poverty.

AN IMPERFECT REASONING

In recent decades, the economic model has vastly expanded the production of industrial goods and commodities with the use, on a scale never seen before, of fossil energy sources. There are those who argue they are emissions for survival in contrast to the luxury emissions of rich countries that benefit from two hundred years of industrialization. But these arguments are merely figures of speech. We are experiencing the end of an economic model due to the impracticality of continuing to increase energy production through fossil fuels.

The crisis of 2008/2009 clearly demonstrated what we had already been watching. There is a lack of limits on the activities of financial agents. The financiers and controllers of international capital flows have never had so much freedom, going in and out of countries according to their interests. Social dumping (low wages?) and environmental dumping (absence of regulations?) are practices that some countries – e.g. China – use to lower their market prices.

British historian Tony Judt, in his last book published during his lifetime, *Ill Fares the Land*, gave a lucid review of the moment in which we live: "(...) markets have a natural disposition to favour needs and wants that can be reduced to commercial criteria or economic measurement. If you can sell it or buy it, then it is quantifiable and we can assess its contribution to (quantitative) measures of collective well-being. But what of those goods which humans have always valued but which do not lend themselves to quantification? (...) Markets do not automatically generate trust, cooperation or collective action for the common good. (...) No one today could claim with a straight face that anything remains of the so-called 'efficient market hypothesis' (...) Our disability is discursive: we simply do not know how to talk about these things any more. (...)

we cannot continue to evaluate our world and the choices we make in a moral vacuum."[4]

The state is an important agent in modern economies. Many processes of industrialization only occurred, historically, with the decisive support of government agencies. But sometimes, the state exaggerates its hand, insisting on development projects that do not meet specific environmental demands, trying to change traditional economic vocations and secular cultures, as in the project of the Belo Monte Dam in the Brazilian Amazon.

I saw firsthand the tragic case of an economic development project funded by the state, which generated unrecoverable environmental liabilities: the disaster in the Aral Sea, known as the "Quiet Chernobyl".

THE QUIET CHERNOBYL

The Cold War caused a competition between antagonistic systems not only ideological, but also in productivity assessments. Regardless, none of the ideologies worried about the environment. Destroying to build and enjoy *as long as it lasts* stood above any models.

In the 1990s, I went to the Aral Sea in Uzbekistan, Central Asia at the invitation of the World Bank, to help with a diagnosis of a possible way to recover the region. It was sad to see... fishermen's trawlers kilometres away from the salty margins of the sea, which were withdrawing more and more.

The Aral Sea was a huge salt water lake with over a thousand islands in its bed, and seemed to be an inexhaustible source of water. In spite of being a lake, it was so big that it was called a sea. In 2007, it had reduced to 10% of its original size and to this day it is still advancing in an irreversible process of desertification.

The death of the Aral Sea is a typical example of the failure of a political-economic model; a model created by Stalin and then followed by Kruschev. Moscow launched the project in the Aral Sea in 1960s.

Entire orchards were uprooted to make way for cotton plantations in the Aral. The desire to expand cotton production increased the dependence of Central Asia on irrigation, especially Uzbekistan. One of the greatest environmental degradations of the twentieth century happened there; a quiet degradation that took shape slowly and imperceptibly over forty years.

James Wolfensohn, World Bank President when I received the invitation to visit the Aral Sea, is one of my long-time fellow travellers.

Below, I present our dialogue in 2011:

CONVERSATION WITH JAMES WOLFENSOHN

Hi Jimmy! You've done so many things in your life. I remember when you were elected president of the World Bank. We were having lunch together here in Rio, in the restaurant of Hotel Caesar Park, if I remember correctly. The phone rang. It was news of your nomination. When you took the position, I suggested that you meet with environmentalists. We had dinner with Tom Lovejoy, Mohammed el-Ashari and our dear friend Maurice Strong. The outcome of this meeting was that you installed Maurice in an office next to yours, so that he could advise you on environmental matters.

Now I ask you: after these 10 years of experience at the head of the World Bank and your return to life as a private banker, how do you see the future of this economic model, which even now has not included the validity of the environmental impacts caused by humankind?

Wolfensohn – I believe that the business community and governments are now becoming more aware of the crucial need to include the environment in basic investment decisions. Not only has there been global political pressure but I believe that the leaders now recognize that the environment is not only an issue for specialists but of crucial importance to everybody. You well know the impact of global warming and how it affects our air, our water, our atmosphere and the temperature in which we live. And of course it has a profound impact on the way we live. These are challenges that can only be met by including environmental investment along with economic decision making.

– Your son, Adam, is deeply involved in the mechanisms referring to the solution for the problems of climate change. Your son knows better than we do all the cap-and-trade mechanism; he knows the difficulties and also knows that the mechanisms today are exhausted, since the market solutions from Kyoto have done nothing to reduce emissions. How do you see the solutions for greenhouse gas emissions now that the current panorama is not functioning?

Wolfensohn – Very sadly, there are many good ideas but we do not yet have a global commitment to mechanisms that will address the issues that you describe. We know that it is possible to follow some of the cap-and-trade mechanisms that you describe, but in the current climate of economic difficulty, I fear that the environment has been put back in terms of priority. I am optimistic that our leaders understand that issues, but I am pessimistic on the timing when they will seriously take up the environment as the critical basic issue.

– We are both well aware that the necessary alterations to the governance models, both of companies and the State itself, have to change. They have to change because they are inefficient with regard to the expectations of a growing global population, of globalization in the broadest sense of the word, in which there are no longer ghettos where information does not penetrate and where collective participation at all levels is increasing even more.

In your opinion, what would be the ideal governance model that is at the same time pragmatic in relation to nations and finalistic in relation to the consequences of globalization?

Wolfensohn – Very sadly, the issue, I believe, is not one of governance. I believe it is one of information and of priorities. Leaders of government and leaders of corporations seem to address short term issues of either re-election or profit, depending on their positions. Longer term issues such as the environment and even globalization tend to get relegated to the second rank. Even in the United States today when one talks with businessmen or with government leaders, the interest tends to be in the relations with the OECD countries – with the richer countries. Of course there is recognition that by 2050 more than half the world's income will be generated in Asia, but this is an intellectual understanding and I feel that our current crop of leaders are looking short term rather than long term. It is to be hoped that the next generation will be more global in their thinking and more realistic in their assessment of the type of world in which they will live. There is no time to waste.

WHAT BEING SECURE MEANS

For my generation, World War II was seen to be last war. We wanted to believe that this time humanity had finally learned the lesson. But with the onset of the Cold War, a new arms race began. And not between European powers anymore, but between two world superpowers. And this time, nuclear powers. Under the impact of the Cold War and obsessed with the dangers of nuclear war I wrote, in 1966:[5]

Of course there was, in the race of the great powers, in terms of economic performance, an obvious gain in speed in the development of liberal democratic societies. But one cannot fail to note a parallel between the structures of production of socialist and democratic superpowers, which become increasingly capital-intensive and less connected with human issues. This can also be extended to defence systems, forcing Western countries to review their systems of production, which is somehow paradoxical to their own economic theories.

The Cold War tension reached its pinnacle in 1962, with the Cuban Missile Crisis. It was just when I completed my internship at the White House, in Alliance for Progress. There I went, invited by Adolf Berle, former U.S. Ambassador to Brazil during the war and named by Kennedy first secretary of the newly created Latin American Desk for the U.S. State Department. I watched all those events unfold.

An enormous amount of global resources is spent on armaments, which governments often call defence. Military budgets distort finances and the global economy, reversing priorities.

Following the example of the United States and the Soviet Union, new countries which have become emancipated throughout the twentieth century invested heavily in their militaries, aiming at regional power and wider influence on the international stage. The old British Raj, for example, was divided into two countries at the time of its independence (India and Pakistan) that became nuclear powers less than thirty years later.

The United States remains the number one in military spending, almost surpassing all others together. Wars like the Korean war and Vietnam war kept the military budgets high in the 1950s, 1960s and 1970s. Recent wars in which the Americans have been involved – Iraq and Afghanistan – increased the already huge budget by about 60% in the decade of 2000–2010, especially after the attacks of September 11,

2001. The nuclear escalation of Iran is one more encouragement to keep resources flowing to the already bloated military budgets.

It is good to have an idea of the magnitude of these numbers. The data are from SIPRI (Stockholm International Peace Research Institute), the influential Swedish think tank dedicated to this subject since 1966. Swedish institutions tend to be independent. According to the latest survey conducted by SIPRI, from consolidated data from 2008 and 2009, we observe that the world spends about $1.5 trillion a year on armaments and military forces. The United States alone accounts for $650 billion, or about 42% of the total. This means that only one country represents almost half of all military spending in the world, taking trade-offs into account. At the same time, this is what maintains American power, which has military bases around the world.

Not even with all this amount of weapons has the world become safer. On the contrary: when we compare the current scenario with only ten years ago, we see greater concentrations of political, economic, and consequently, environmental instability. The so-called War on Terror, launched after the attacks of September 11, 2001, alone has already consumed about $1.3 trillion in a decade of wars. The result has been a less secure world. Iraq was devastated; a strengthened theocratic Iran; Afghanistan becoming once again the world's largest opium producer, and Pakistan increasingly unstable due to the presence of Taliban Islamic extremists. How secure is that?

But not only the United States insist on this game. Other countries that spend a lot on weapons are China ($85 billion), France and Britain ($65 billion each) and Russia ($58 billion). All of them increased their military spending between 2000 and 2010. The only two that decreased their military spending over the last ten years were Germany (which spent $50.8 billion in 2002 and 46.8 billion in 2009) and Japan (from 48.5 billion in 2002 to $46.8 billion in 2009), precisely the powers which were defeated in the Second World War. *Et pour cause?*

The memory of the militarist adventure caused traumatic damage to the population; that is perhaps the reason for this downward trend. Still they are the sixth and seventh countries which spend the most in this area.

This immense volume of money is named "defence spending", but it actually means an investment that aims toward death, not life, because it is a bet on war and on political instability as being likely scenarios. This could be a self-fulfilling prophecy. The numbers are totally disproportionate when you consider relevant demands: fighting environmental degradation, investments in education, sanitation... areas which could induce sustainable development and bring real safety to citizens, that is, a sense of safety in a broader sense. We can take this reasoning even further.

ENVIRONMENTAL PROBLEMS AND ARMED CONFLICTS

In the future, armed conflicts will tend to be caused by environmental problems. The new geopolitics will be based on assuring a supply of basic commodities. The amount spent on military budgets in 2011 is equivalent to or greater than the income of the poorest half of world population.

The question that arises is: can we obtain more resources, currently allocated to military budgets, for items on the environmental agenda? Definitely, yes, since each expense on military activities decreases the amount of resources available for the development of clean technologies. These investments should be the highest priority in national budgets, bearing in mind that both the solution of environmental problems and future economic models will be based on innovation.

Norman Myers, in his book *Ultimate Security – The Environmental Basis of Political Stability*, mentions some examples of trade-offs related to this dilemma. The world's military spending is about $4 billion a day, according to current data from SIPRI. To reverse the process of desertification under way in several places of the planet, we would have to spend $12 billion, that is, three days of military spending. To bring clean water and sanitation to one third of the world population, currently excluded from these services, we would spend $36 billion per year, or nine days of military spending. To boost the agriculture sufficiently to totally eliminate hunger, we would have to spend about $40 billion in one year, or less than 10% of the U.S. spending on defence. We could go even further in this reasoning to reach the same conclusion: with a tiny percentage of the resources spent on militarism it is possible to solve many problems. Social and environmental issues go hand-in-hand. Which one offers more security: missile-equipped high-tech stealth aircrafts or a world population that is economically included?

Myers also quotes Maurice Strong. According to Strong "Earth Incorporated is literally in liquidation. Much of the income we are producing isn't really income at all: it's running down our capital. If we used only 2% of the global GDP, we could run the Earth on a sustainable basis." Myers estimates that the cost of solving the major environmental problems is about $700 billion a year or about half of the world's military spending. That is, a 50% cut would already enable a huge flow of resources.[6]

I will always remember a quote from Willy Brandt, in the 1980s: "The arms race is killing people without using weapons."

Lester Brown, author of the "Plan B" series,[7] offers a more modest calculation. According to him, about 13% of the resources allocated to military budgets would be enough to give a boost on the path of sustainable development, together with other economic and institutional actions we will discuss later.

NEVER UNDERESTIMATE THE POWER OF IDEAS

The Nobel Peace Prize of 2004 awarded the future and criticized the present. It posed a welcome question: why did activism for sustainable development receive the Nobel Peace Prize? Wangari Muta Maathai is a black woman, a biologist, humanist, and, above all, she believes the human presence on the planet is linked to the basic principles of social justice, respect for human dignity and most importantly the preservation of natural resources, a gift we received at the dawn of time. This African woman represents all of us who, in the last twenty years, sought ways for a new model to give us hope for the future.

I remember Muta Maathai participating in the Rio+5 meeting in 1997, when we confirmed the necessity of adding the social factor to the economic vector of sustainability, as an end recipient of this development. Why was only Wangari awarded the Nobel Peace Prize, and not all those who believe in the need to review the political and economic model? Why, after the pioneer Gro Brundtland, another woman, this time African, should receive worldwide recognition for her actions in preservation of the environment and social justice? Wangari worked on the fields, studied at university, participated in politics and understood that the world is not only the territory in which she lives. The Nobel Prize she received extends beyond Kenya to the rest of the planet.

I do accept Wangari's call to form a tribe whose mission is to add elements of sustainability to the political, social and economic model, which is only possible through enlightened leadership. The presence of guiding voices is important, but the transition from one model to another depends equally on the evolution of science, on the development of public awareness through education, and on the integration of public policies with techniques of corporate sustainability.

Over the past fifty years, the planet went through wars, conflicting ideologies, demographic explosion, and environmental destruction. The world is still hanging on the idea that security is only achieved through military forces.

Billions are figures outside palpable reality. Whatever the ideal amount to start the changes, the most important thing is to change our mentalities. How can we persuade people about urgent action?

Strategic decisions usually take place in silent offices. Obviously, opposing the military-industrial complex is an enormous challenge. Non-opposition is the same as letting this subsystem of the world economy grow more and more, with its destabilizing effects.

There is an absurd inversion of priorities. If we think of two areas in which humanity currently spends the most, we will come to a sad conclusion. Our priorities are weapons and drugs. The fact is that the military and oil and industrial complex dominates the globe. Neither Napoleon, nor the British Empire, nor the United States, each in their own time and dimensions, managed to achieve that.

GROSS DOMESTIC PRODUCT, THIS OMNIPRESENT STATISTIC

GDP is so widespread a concept that it has become almost a noun in the routine of news listeners and readers, but little is known about its principles or about what this concept measures and means.

Gross Domestic Product is the measure of the totality of economic activity of a country, indicating its aggregate level of activity or production. Most economic policies are based on this measure and on its growth.

When the Second World War ended, the GDP began to be used as an instrument for economic policies, guiding even the policies of the newly created IBRD and International Monetary Fund, as well as the global financial markets.

The GDP measures the total production of a particular country – in agriculture, industry and services. Economic growth is measured through the rate of GDP growth and, mainly, through the rate of real GDP *per capita* – that is, by discounting inflation and population growth. Below, we will see what the GDP excludes.

WHAT THE GDP DOES NOT MEASURE

Certain services are not properly included in the national accountancy, as is the case of informal activities. Other services, particularly financial – which keep themselves *a quo* of other systems of regulation – are overestimated. Corruption is another factor that distorts the national accountancy data. The bigger the State machine, the more likely the existence of corruption. There is no measure of the quality of education, nor of access to a public health system, nor of income inequality or the deleterious effects of lacking infrastructure in big cities.

Productive processes generate by-products that are not accounted for by the market.

Since fuel consumption increases ratings of GDP, traffic jams give us the impression that societies are richer. Seriously worrying externalities, such as pollution and emissions of greenhouse gases cause collective damages, none of which are measured.

The GDP is a symbol of progress, but it is far from the complexity of economic reality and does not reflect values of social well-being or environmental damage.

We need new indicators and parameters for an economic model that meets the tripod of sustainability – economic, social and environmental. Many thinkers have reflected on possible new indicators, and are converging toward the idea that the pillars of a new economy should have foundations going much further than the measurement of production, consumption and services. We are at the beginning of a new path, although still walking very timidly.

WHAT NEGATIVE COSTS ARE

The empirical basis of GDP has never undergone thorough review. In 2008, the French government commissioned a report on new economic indicators. The Sarkozy Commission, as it became known, focused on three topics: 1) Issues relating to the current GDP; 2) Quality of life; and 3) Sustainable development and environment.

In relation to issues related to the GDP, the Commission recommends some adjustments in order to make the index express some economic, social and environmental realities. There should be a transparent attempt to evaluate and price the environmental assets and services of a country, so that if one habitat is degraded, the depreciation of natural capital is accounted for as losses.

Another recommendation is that certain expenses should not count as produced wealth, because they do not directly benefit the population. These are negative expenses, which do not add value to society and may result in social or environmental deterioration. For governments, examples of negative expenses are the military sector, the cost of prisons, and cleaning up ecological disasters. For individuals, useless expenses are the losses caused by traffic jams, such as wasted time and fuel.

One of the weaknesses of the current measurement of GDP is that services provided outside the market are not considered. Housework, for example, is only accounted for when there is hired labour for those tasks. The same work done informally by a family member does not count as wealth.

But how can we properly measure something as impalpable as the well-being of a population? Quality of life is a much broader concept than the conventional categories of economics. Wealth can be transformed into well-being in several ways with more or less success, by different peoples and cultures, and this depends on political and speculative arrangements. According to the Sarkozy Report, in the United States, the total spending on health (individual and state expenditures included) is the highest in the world. However, the health of the American population is on average worse than the health of Europeans, who spend much less in this area. There is a clear discrepancy between *input* (health expenditures) and *output* (effective health of the population), mediated by factors such as lifestyle and diet.

Quality of life is guaranteed both by material and non-material resources, and it is in this latter that the difficulty of measurement resides. But difficulty does not mean impossibility.

This is an interdisciplinary field. Can we measure subjective well-being using questionnaires borrowed from psychology? Could we also have an approximate measure of functional abilities (how individuals are integrated in their social environment) and relative freedoms (the options and choices available to them)? Can we also try to establish a way of measuring happiness, even if this requires philosophical questions and if its definition is necessarily subjective?

These questions led the UN, in 1990 – long before the Sarkozy Report – to create a new index, the HDI (Human Development Index), including education and health on the side of production. But regarding environmental damage, the silence from statistical authorities is striking.

The big question now goes beyond: how can we seek ways to measure not only what is traditionally understood as well-being, but also the costs of environmental destruction? The Sarkozy Commission still considers the HDI an insufficient index, because part of it is tied to the GDP.

The challenge continues as we give relative weight to each one of the items comprising quality of life: general health, ensured freedoms, level of education and skills, personal activities outside of work, good urban conditions, social connections and finally, environmental conditions... which takes us to the third and last item of the agenda proposed by the Commission: what does a measurement of sustainability mean?

SUSTAINABILITY?

Since the Brundtland Report – Our Common Future (1987) – the notion of sustainable development has been steadily broadening and is today a concept embracing all economic, social and environmental dimensions in the present and in the future. For this reason, it is not possible to achieve a single index to measure sustainability. There is no magical number to be followed. Indexes like the Green GDP are insufficient. Uncertainty itself and the evolution of scientific knowledge on the subject lead to a hybrid approach.

The recommendation of the Sarkozy Report is to use a panel of indicators separated by more easily measurable items, with levels of atmospheric and water pollution, habitat destruction, marine resources, and mineral resources. These factors must be taken into account. Natural capital reserves, renewable or non-renewable, should be measured, as well as variations in annual reserves. These data, combined with other economic and social indicators, could provide a more accurate picture of reality.

This is not the same as replacing the GDP index with a more ecological index. It is more like seeing the multiplicity of factors that make up life; the complexity of human interference within the environment in which we live.

WE ARE MAKING THE WRONG CALCULATIONS

Countries strongly based on non-renewable resources make more money on the rapid exploitation of those resources. However, once depleted they no longer generate wealth. An example: oil still unexplored is not considered in the GDP calculation; only the amount extracted per year enters the statistics. The same goes for forests.

Environmental services provided by forests (soil quality, carbon sequestration or protection of water resources) guarantee life but do not generate market value and are not included in the GDP calculation. On the other hand, wood that is sold on the market is included.

Robert Kennedy, in a speech at the University of Kansas in the historic year of 1968, drew a good picture of the importance of measuring what is valuable: "Gross National Product counts air pollution and cigarette advertising, and ambulances to clear our highways of carnage. It counts special locks for our doors and the jails for the people who break them. It counts the destruction of the redwood and the loss of our natural wonder in chaotic sprawl. It counts napalm and counts nuclear warheads [...]. Yet the gross national product does not allow for the health of our children, the quality of their education or the joy of their play. It does not include the beauty of our poetry or the strength of our marriages, the intelligence of our public debate or the integrity of our public officials. [...]. It measures everything, in short, except that which makes life worthwhile."[8]

THE ARCHITECTURE OF THE NEW ECONOMY

In 2010, an academic group produced a report on climate policy called the Hartwell Paper. The aim was to reassess climate policies after the crisis of 2008/2009. The document was a partnership of the Institute for Science, Innovation and Society, Oxford University, with the LSE (London School of Economics).

The Hartwell meeting was held under the Chatham House rule, in which the identities of participants are not disclosed but they assume joint responsibility in the final result. In its executive summary the report states: *"The crash of 2009 presents an immense opportunity to set climate policy free to fly at last."*

The Chatham House Rule is formulated as follows: when a meeting (or part of the members of a meeting) is governed by the Chatham House Rule, participants are free to use the information received, but are not allowed to disclose the identities and affiliations of the speakers and participants.

This rule was created in the 1920s by the Royal Institute of International Affairs – Chatham House (England), to encourage free debate and information sharing. It is used to ensure anonymity. During the discussions, one can not reveal who said what, to protect individuals from any censorship or pressure in the future.

In the final report of an event, all participants must collectively sign, without highlighting any individual participation or ideas.

This rule underwent its last reformulation in 2002, and its developers have paid attention to recent innovations in communications. A participant may report via Twitter what is being debated, under the condition of respecting the confidentiality of the names of debaters. The scene of the Hartwell Paper was Hartwell House, in Buckinghamshire, England, once a refuge for the secrets of Marie Joséphine, wife of Louis XVIII.

A GOOD CRISIS SHOULD NOT BE WASTED

The financial crisis of 2008/2009 has left us a window of opportunity for change. One of the points of advice provided by the Hartwell Paper is that a good crisis should never be wasted.

The Hartwell Paper analysis criticized the way in which climate policies were understood and practiced under the umbrella of the Kyoto Protocol. The conclusion is that these policies have failed and did not produce any reduction in emissions of greenhouse gases over the last fifteen years. The Conferences of Copenhagen (2009) and Cancun (2010) approved documents of dubious status with unclear commitments and consequences.

According to this paper, the reason for the inefficiency of the Climate Convention/Kyoto Protocol approach is that it systematically failed to understand the nature of climate change as a political issue.

Climate change is the result of an open system, which includes several vectors, such as the economy, politics, and their cultural constraints.

The Hartwell Paper proposes a positive agenda, that is, an organizing principle for environmental efforts emphasizing well-being by way of three broad goals: 1) Clean and cheap energy for all; 2) Economic development that does not compromise natural systems; and 3) Adaptation for extreme weather events, regardless of their causes. For this purpose, the energy matrix must be decarbonised, which leads to the need for investments in innovation and subsidies for clean energy sources. This funding could arise from a fund formed by progressive carbon taxation.

The report shows that, as no agreement could actually reduce emissions of greenhouse gases, the trading mechanisms of multilateral diplomacy also began to be put into question.

The Kyoto Protocol had great historical importance, but its market mechanisms have proven to be limited. As we have seen, the U.S. has not ratified it, and China is not obliged to reduce the emissions of greenhouse gases. Between 1990 and 2008, the United States had an increase of 14% in emissions of GHG. If they had ratified the Kyoto Protocol, they would have had to reduce 7% of emissions. They not only failed to make any reduction, they in fact made an increase. This leads us to a gap of 21% in relation to the goal established by the Protocol. China, in turn, which in 1997 had no commitment to reduce emissions, had a 191% increase in GHG emissions between 1990 and 2008. China and the United States account today for practically half of all annual emissions of greenhouse gases.

Despite the establishment of targets for developed countries and the creation of several mechanisms, among them the CDM (Clean Development Mechanism), emissions have increased continually over the last twenty years. Kyoto was the first treaty with targets and timetables but, in spite of being legally binding, there are virtually no enforcement mechanisms. Although there is no clear path on how to reduce emissions, we have one certainty: the tools we have are not enough. The solution we expected from Kyoto did not take place, as shown in the following table.

Year	**CO_2 global emissions** from fossil fuels, cement production and land use change[9] (billions of tons of CO_2)	**Variation (after 1990)**
1990	27.8	–
1995	28.9	+ 3.9%
2000	29.9	+ 7.5%
2005	32.9	+ 18.0%
2009	34.0	+ 22.3%

RICH AND POOR

I do not agree with the commonplace thesis that developing countries have a moral right to pollution. Climate change is a perverse problem, not easily identifiable in its effects. Its complexity is not reducible to simple models, and its causes are rooted in the social and economic systems we adopt.

It is a utopia to think that there is one single scientific reason, which could then be used through political consensus to resolve the issue. There is still a lot to be studied by climate science, but there are already certainties. Slowness to act comes from political drag, not via scientific means; it originates in political and financial interests. What they want is to gain time. How much time do we have left? What is the schedule of disaster?

Of course we also should not disclaim the responsibility of informed or uninformed consumers who do not demand their rights either for themselves or for their descendants.

Returning to the Hartwell Paper: I was very excited about the report, whose proposals reflect my own ideas. The central thesis is carbon taxation, initially small but progressive, which would form a global fund linked to innovation and the development of clean technologies. It is similar to the proposed elaboration of the Clean Development Fund that we took to Kyoto in 1997. I believe that only with this mechanism will it be possible to provide cheap and clean energy to humanity – it is not small potatoes, but it is possible.

The whole problem of climate change is understood as a consequence of the effects of a highly complex system, which indicates no solution will be effective if it comes from a desire to maintain the foundations of that same system. Huge economic interests, very powerful companies of an economy driven by intensive carbon use... we experience and build our world with the feeling that natural resources are endless, but we need always remember that it is not the planet that is under threat, but our own species. The change will happen. The question is whether we will lead this process or be forced to act or perish, based on the signals nature has already shown us.

SUSTAINABILITY AND BUSINESS

"Is it progress if a cannibal uses a fork?" John Elkington, in *Cannibals with Forks – The triple bottom line of 21st century business* brings us these words from Polish poet Stanislaw Lec to illustrate his concept of corporate sustainability.[10]

The image is that of companies as cannibals, swallowing one another in mergers, acquisitions, and global restructuring. Can cannibals be civilized? The fork symbolizes increasing awareness, which can be seen as the practice of sustainable ethics.

One of the pillars of our work in FBDS takes place in the field of corporate sustainability. What does that mean? It means taking indicators and instruments to companies in order to make them adopt practices consistent with sustainable principles. A key point in our dialogue with companies is to show that sustainability aims for a future vision. We organize frequent workshops which function as incubators of concepts and actions.

Since the early days of FBDS, I thought it fitting to strengthen partnerships with various research institutions to conduct seminars, which have become routine practices in our activities. In one of these workshops in 1996, about "Business Opportunities and Short and Long-Term Risks" I reasoned that

sustainable development looks towards the future. The financial market has short-sighted concerns. Accounting and financial reports do not take into consideration environmental risks, depletion of resources and the opportunities that can be generated by environmental capital. The investment market makes decisions based on incomplete information.[11]

FBDS, in its development of propositions and actions, is currently a mix of consultancy and think tank. We want to strengthen the ideas and practices of companies for the productive sector to take on responsibilities that go far beyond accountability to shareholders.

Companies must not only create value for stockholders (shareholders), but to stakeholders, in other words, all interested parties directly or indirectly affected by business activities. Social and environmental assessments should be transparent in all risk assessments.

FBDS has been working to call and to meet with companies through debates and the production of reference studies. We work daily to show companies they need to include the sustainability agenda in their business strategies, and not only in their marketing and communication speeches.

We already perceive a change in perception, but this is still insufficient. It is necessary to define stronger corporate policies, including an environmental ethics code. Many companies have already been touched on their periphery by the environmental issue, but not yet in their hearts.

How to reconcile pressures for profit and short-term results with the maintenance of long-term natural capital? How can we unite business opportunities and consistent social and environmental practices that do not harm the competitiveness of companies in the global arena? This is the great challenge, and we can begin to sketch some answers. One way is to implement self-regulating industry schemes with strong regulations.

CORPORATE SUSTAINABILITY IS POSSIBLE

Sustainability indexes and reports are still imperfect and therefore questionable tools. For example, one of the companies that appeared at the top of the Dow Jones Sustainability Index, for its innovations in sustainability, was British Petroleum. We expect that the accident involving BP in the Gulf of Mexico in June 2010 has deeply touched the issue of sustainability in businesses.

Their shareholders were not the only ones who strongly felt the blow. Where once there was a risk of a major accident, this potential was not detected by the forms of the Dow Jones Sustainability Index.

In practice, the tools still do not work as intended. One of the commitments made by companies is the eradication of child labour. But do they really check on the actual working conditions of their suppliers?

WHO PAYS THE BILL

We still have another level of difficulty. It is not yet known how much this new business model will cost. Often consumers are not willing to pay more for a product or service coming from sustainable practices. These practices are still more expensive. Nor do shareholders want to cover the expenses, which compromise dividends. The board does not decide either, and the company does little or nothing. Who should pay for the necessary changes?

To date, there are no higher returns for those who invest in companies listed on the ISE (Corporate Sustainability Index) of BM&FBOVESPA. The average rate of yield is the same for those who invest in the portfolios of traditional companies.

The problem is a lack of pricing for environmental and social externalities. Companies are still not able to measure the value of sustainable practices. If the economy does not change, what will the cost be in the near future? Companies that now adjust to something that will only be regulated in ten or twenty years would be betting on the future, but competition is established in the present. Governments could accelerate this agenda. If there is regulation, the gears may work. Otherwise it becomes voluntarism.

Could regulation and pricing of environmental components become a reality through multilateral agencies like the UN, the World Bank, IDB and all financial agencies, regional and global? I don't think so. Perhaps this topic should be picked up by the WTO, when discussions between countries would be made according to product, with international trade guidelines, and not between diplomacies in the Climate Conferences, that is, between national policies.

In Brazil, sectors already regulated such as the electric power sector have better sustainability practices. Fines are now priced losses, that is, placed in the present, not in an abstract future.

Public awareness is on the rise and there is increased vigilance from the media. Both are strong engines to help push the process. But the sense of long-term future is something very abstract for the human mind.

Companies know how much it costs to create an image and how much it can cost to see it destroyed. For this reason environmentally disastrous actions are treated very carefully.

What are the key measures a company should adopt in the process of implementation of sustainability? We have returned to our triple bottom line. Social, economic, and environmental ideas form a new conceptual universe that must enter the business plans of the twenty-first century: energy efficiency, water management, conservation of biodiversity, impact on land and habitat, control of the emission of greenhouse gases (GHG), solid waste management, actions against child labour, slavery, and corruption, emphasis on community education, and transparency in the disclosure of economic flows.

POLITICALLY UNFEASIBLE?

It was August 31, 2010. The day before we had held a meeting with bankers at the Foundation.

This meeting made me feel frustrated. Scattered excerpts of thoughts... this is the economic-financial model that is not willing to give anything up. I have met this resistance for decades. I shall explain this meeting:

A UN study group connected to fundraising had come to the Foundation to seek advice on further actions in regard to climate change. Besides my staff, also attending the meeting were representatives from the Ministry of the Environment. The meeting was an example of the scenario that is repeated in several international negotiations, whether within or outside the UN. Again, I reflected on some issues of financial regulation and political opposition to the proposed changes.

We still think of fundraising without touching the financial-economic model. The great deadlock of that meeting at the Foundation was the difficulty in convincing the financial agents about the necessity and urgency of the changes. We raised the possibility of taxation of only 1% or 2% on the price of an oil barrel, which would be enough to start an investment fund for clean energy, but we heard that "this is politically unfeasible."

While we are not able to convince that *we* and *they* have a common problem, there will be no interest in seeking solutions. We suggest two main sources of taxation: for the richest countries and for the ones who are depleting non-renewable resources (oil, gas and coal). It would be a kind of toll; the greater the use, the higher the taxation.

Is it politically unfeasible or economically inconvenient? Fundraisers, through various instruments, are invested in maintaining the current power system.

Meetings like that are always trying to seek mechanisms to raise the necessary funds in order to amortize the effects of climate change, within the *polluter pays system*. These mechanisms helped many countries maintain their emissions and mobilize resources on which they achieved and still achieve high profitability. So why change? They worry about the so-called unfeasibility of the model as a whole.

What is said to be "politically unfeasible" may, one day, cripple the lives of future generations.

An English friend asked me where to invest in Brazil, and it was not a difficult question. Commodities will have permanent value. But there are still those who want to maximize their ability to use global resources in order to maintain the current financial machine until its exhaustion. *Let's keep the cake and eat it too.*

HOW TO THINK OF A NEW ECONOMY

Some economists are already rethinking basic concepts based on the limitations environmental problems impose us today. However, the real economy still functions according to the same standards of high carbon usage – the business-as-usual situation.

Nicholas Stern's proposal for renewing economic thought is notorious. His study focused on the impacts and risks arising from climate change and the opportunities and costs associated with actions for a low carbon global economy. The Stern Review (2006) touched the most sensitive spot of the financial system: the profit motive. "The benefits of strong, early action on climate change outweigh the costs." The costs of inaction would be about 20% of GDP.

In 2010, I was invited to write the preface of the Brazilian edition of Stern's book, *A Blueprint for a Safer Planet*, in which I highlighted:

Stern's thoughts on public policies relating to climate change led to the creation of tools to achieve new models in economic and social sciences. The instruments for prediction created by the IPCC induced Nicholas Stern to format economic tools that clearly demonstrated the cost-effectiveness of implementing new public policies, both globally and in Brazil. (...) Whereas the problems intensify with globalization, Stern's work represents a thought where the international perspective lies in the technical and scientific view of the consequences that will result if the remedies needed to decarbonise the energy matrix are not implemented.[12]

Stern's career enabled him to propose a shift in economic thinking. Chief Economist of the World Bank between 2000 and 2003, Stern was invited by the British government to review work on the economics of climate change. It is interesting that Stern was not an environmentalist but a scholar, respected for his work with poverty reduction.

The *Stern Review* of 2006 had a huge reception in the media and elite political circles, clearly explaining the risks of inaction and the benefits of action, establishing correlations between demographic and economic growth, increased emission of greenhouse gases, carbon stocks in the atmosphere, increase in global temperature, and climate change resulting from this process. To avoid irreversible and disastrous climate change, the main goal is to keep emissions at a level that can be absorbed by the planet. But how to do it in a world whose population and economy will only grow?

Stern argues that economists should contribute to this discussion by examining the decisions or the ethical policies of economic alternatives and their consequences.

However, there is little room for ethics in the formulation of economic models. Consumption is still the main indicator of social progress, along with the concept of social utility.

In 1992, the year of the Earth Summit in Rio, I already observed that *to become reality, the proposal for sustainable development requires that rich countries change their economic model and give access to countries that still have not reached the minimum levels of income, technology and resources consistent with their basic needs. (...) We all know there are great economic and financial costs to be paid in order to correct directions and past mistakes. However, none will be heavier than the consequences we will all suffer in case we are not able to accomplish the goals that science and common sense tell us to. (...) The sustainable development proposal is essentially ethical, and will only become reality if the ethical parameters are emphasized and considered a priority in the way they deserve.*[13]

Inspired by the Stern report, environmentalists from different areas of expertise met to produce the study "Economia da Mudança Climática no Brasil: Custos e Oportunidades" ["Economics of Climate Change in Brazil: Costs and Opportunities"],[14] with support from several academic centres[15] and the Ministry of the Environment. The result was a comprehensive economic analysis of the problem of climate change, classified by subject and indicating regional differences in Brazil. It was a pioneering study and an opening to broader research lines which could strengthen every aspect of the subject.

The problems presented in the report are of great importance to the Brazilian development agenda. Mitigation, adaptation, and management are the ways to face the problem, but they depend on the mobilization of society and its technical and political sectors. The alternative is to face a "framework of significant impact of climate change on biodiversity, ecosystems and the services they provide. The losses would be around $26 billion per year. (...) The estimate of the material values at risk along the Brazilian coast is of 136 billion to 207 billion reais."[16]

FBDS was able to make a contribution to this broad study with its work on water resources.[17] Preliminary results are alarming. Throughout Brazil the trend is of a decrease in river flows, including regions where the models indicate increased rainfall, because higher temperatures will cause higher evapotranspiration. In the semi-arid Northeast, the data suggest that in just one or two decades, the situation may be critical, with intense draughts, internal migrations, worsening of socioeconomic conditions, and increased demand for services in urban areas of the region. We recommended strategic planning and integrated management programmes for the river basins and special attention to areas and populations at greater risk.

Therefore, we must integrate economic concepts into a broader view, both from a spatial point of view (thinking about Earth as a unit) and a temporal point of view (taking care of the rights and resources of future generations). Of what use are economic development projects that do not take into account these issues?

It is very likely that future generations will have more technological knowledge and wealth than us, but, in contrast, they would have fewer natural resources, which would have been consumed by us and our ancestors.

It is not fair to sacrifice the well-being of future generations in favour of our own. Men tend to value what is near more than what is distant. In this sense, would future generations, which are not of our time, be of less value? Today, this issue is collective in the most broad sense, since it is about the survival of mankind.

Gro Brundtland brought up interesting thoughts when she said that "future generations neither vote nor pay taxes, thus they are not of priority to current leaders."

We are going through transformations in our assumptions of conventional models. Thus, many of the basic concepts of traditional economics cannot simply be transposed onto the issue of climate change.

Environmental goods and services will be more valuable in the future than today, in regard to their scarcity and deterioration. The next generations will be better off with a better environment, even if that means less economic growth.

Often we do not take into account an important fact, largely forewarned by environmental science: the impacts of decisions we take today will only be felt in twenty or thirty years, with consequences throughout the coming centuries. We tend to be short-sighted, looking a maximum of twenty or thirty years ahead. However, we entertain a prospective affection for our posterity. But these markets do not establish priorities of resource allocation on such a long period. Even the largest companies do not plan big projects and investments for a period longer than twenty or thirty years. These are the horizons of market mechanisms. Markets have a short-term vision; medium-term at best.

However, atmospheric cycles and the relationship between the flow and reserves of carbon are interdependent and long-term processes. Markets do not reveal what are the best decisions we should take to ensure that coming generations do not suffer with radical changes in their environment.

What the laws of economics can predict with certainty is related to the principle of scarcity value: if we invest more in goods or services instead of conventional environmental goods and services, in the future, prices of the latter will increase in relation to the former. The flow of GHG accumulates in atmospheric reserves, making a reversal increasingly expensive and complex. The longer we take to address the environmental issue, the higher the costs.

Returning to the Stern Review: it is cheaper to act now, to prevent the problem, than it is to spend in the future to try correcting disasters.

Science gives us evidence of climate change and its anthropogenic causes. If we continue with the business-as-usual model, in the end of the century we will have reached the worst possible scenario: a frightening mark of 750ppm of CO_2-eq and a possible increase of 5ºC in global temperature, what would lead human beings to a completely unknown world in the next century and one probably not appropriate for human life. This is the real crisis.

In our daily lives a variation of 5ºC may not seem very dramatic, but when we refer to the average global temperatures, it is a huge alteration. During the last ice age, when glaciers covered almost the entire planet, the Earth was on average only 5ºC colder than today, and many species today have already become extinct. This is what will happen to us, *homo* and not so *sapiens*.

Since 2006, the year of publication for the *Stern Review*, climate science has advanced, its methods and procedures have become more sophisticated, and we see that the changes and risks were underestimated.

Stern said, in 2010: "Climate science showed the risks are much greater than we had anticipated, and the pace of the effects is much faster. Therefore, we must cut more emissions than I suggested at that time, and this will require investments around 2 to 3 per cent of world GDP."[18]

Gradually, environmental science invades areas traditionally belonging only to economic orthodoxy. The correct valuation of resources, assets, and carbon pricing, for example, are concepts which would be unthinkable a few years ago, but are now beginning to see debate in the academic environment.

ECOSYSTEMS AND BIODIVERSITY: THE VALUE OF A STANDING FOREST

Assigning value to biodiversity is our next frontier. A meeting of Ministers for the Environment in 2007 at Potsdam proposed a study to verify the economic impact of losing global biodiversity. The initiative has evolved into a project called TEEB (The Economics of Ecosystems and Biodiversity), under the auspices of UNEP. Headed by Indian economist Pavan Sukhdev, the study on the economics of ecosystems and biodiversity extends the same logic that has been applied to climate change to understand the economic role of the environment. In other words, this means understanding the cost of not preserving ecosystems and biodiversity on the planet. It also means evaluating the goods and services nature provides us for free, so we can be more aware of the economic impacts of human activity.

It is a mammoth task, but there is still no solid scientific knowledge on the losses caused by the extinction of species. The economic quantification of biodiversity cannot be reduced to a single factor, it is necessary to study each aspect in different ecosystems around the globe.

But how to evaluate prices of irreplaceable natural systems? Will governments, businesses, and society accept paying for goods currently available for free? These are challenges guiding budding research on the pricing of natural capital.

Dominant political and economic models are the main causes of the loss of biodiversity we see today. Population growth magnifies the consequences of this model by converting original habitats into pastures and urban areas. In Brazil, the deforestation of the Amazon for agricultural activity is the main cause of emission of greenhouse gases and loss of biodiversity. These activities contribute almost nothing to increase

Brazilian GDP. The loss of natural capital is not included in the calculation.

The TEEB study report recommends pragmatic actions: to include natural and human capital into economic accounting, to include services and goods produced by nature in the market system, and to change taxation systems (to tax what is taken out of nature instead of taxing production).

The fundamental importance of goods and services coming from habitats is largely unknown by the general public.

Pavan Sukhdev brought me a pleasant surprise. Amid the rationality of the report, he quoted T.S. Eliot, a poet who has accompanied me since my youth. Pavan brought the poem "Burnt Norton" from the *Four Quartets* series: *"Human kind cannot bear very much reality."*

With all the difficulties and complexity inherent in the issue, we must not fail to produce the best estimates, so society can make choices. Recognizing that ecosystems have economic value is already a considerable advancement.

What are the main services assigned to ecosystems? According to the TEEB report, they are of provision, regulation, support, and also cultural. Let's examine these four categories.

Healthy ecosystems are able to provide the essentials for human life: conditions for agriculture (soil fertility), raw materials (construction and fuels), drinking water (conservation of water sources), and raw materials for medicines and new therapies.

For the pharmaceutical industry, genetic resources safeguarded by the ecosystems generate 25% to 50% of a market of $640 billion.[19]

The services regulated by ecosystems are related to their natural functioning: climate control and air quality at the local level; carbon capture and sequestration, mitigating global warming; reduction of extreme draughts and floods; degradation of waste and water treatment; preventing erosion and maintaining soil fertility; pollination of crops; and the biological control of pests and disease vectors. It is estimated that the contribution from pollinating insects alone generates about $190 billion per year to agriculture.[20]

Ecosystems maintain inventories of genetic material that can potentially be useful for scientific and technological development (biotechnology, nanotechnology, biomimetism).

Finally, cultural services include those intangible benefits that people can get when in touch with ecosystems: promoting physical and mental health; ecotourism; aesthetic, artistic and even spiritual appreciation.

The importance of maintaining biodiversity is not limited to these categories, but we can begin to assess and measure the value between maintaining an intact ecosystem, or converting it into pasture, planting areas, or urban concentrations. In addition to ethical considerations, the economic approach can provide quantitative data to help formulate public policies which provide safe signals for economic agents.

Loss of biodiversity is a problem as serious as climate change. Everything is interconnected: the loss of biodiversity contributes to global warming, global warming accelerates the extinction of animals and plants, which in turn ends up compromising ecosystems. One cannot prioritize one issue over another. This is the challenge of environmental issues: we must attack the problem from every side.

We can intuitively understand the intrinsic value of natural areas. In every city, we feel good when we enter a park and are able to breathe cleaner air that is available for free. There is a small flow of value in the green square where children play. On a larger scale, there is a great flow of wealth when we consider the services provided by large tropical and equatorial forests, such as the Amazon rainforest. In this case, developing countries are the ones enabling, through their great natural reserves, the economic growth of the developed world. Carbon capture and sequestration by trees, the maintenance of a gene pool, and the redistribution of atmospheric water vapour are essential to the climatic balance of all countries, and nobody pays for it.

TOWARDS A GREEN ECONOMY

The concept of green economy has been coming out from the scope of environmental experts and is now invading the mainstream. There is a sense of weariness from the multitude of crises.

According to an observation of the UNEP 2011 report (United Nations Environment Programme), "Towards a Green Economy, Pathways to Sustainable Development and Poverty Eradication,"[21] a worrying scenario has accelerated over the last decade: extinction of species, climate change, shortages of food and water resources, crisis in the financial system and in the economies. The demand for energy and food is growing and will keep on growing in the same proportion as population pressures and the upward social mobility of billions of people. Even with all projections for 2050 indicating a pressing need for major investments to meet these demands, little is invested in the necessary areas.

Since we live in a global situation of water stress, and the prices of food have been increasing, at the same time the world is not preparing for a population of 9 to 10 billion in 2050.

Bad policies have been preventing the transition to a green economy. For UNEP, there is a serious mistake in the way capital is allocated: over the last two decades, there have been heavy investments in real estate assets, fossil fuels and financial papers such as derivatives, but relatively little is invested in renewable energy, energy efficiency, public transport, sustainable agriculture, or in the conservation of ecosystems and water resources. The strategy of investors in general seeks rapid accumulation through capital allocation in physical and financial resources, at the expense of the depletion and degradation of natural capital. Reversing those investment strategies and betting on the rise of the green economy is clearly the way to come out of this situation.

The pathway for development lies in the maintenance and increase of natural capital, which should be seen as a critical economic asset and a source of public benefits, especially for the portion of the population with lower incomes.

Two myths need to be overturned. The first is that there is a conflict between environmental sustainability and economic growth. This is false. Green investments are able to provide strong growth. A great example is basic sanitation: it protects nature and health, provides visible development, increases social well-being and also creates jobs and income.

The second myth says that the green economy is a luxury that only rich countries can afford. This is also false. Of course, with more money available, rich countries have resources to implement a new economy, but there is no reason why developing countries would not be able to invest in this transition.

UNEP has made three key findings about the green economic model. Greening the economy generates wealth, especially the wealth represented by natural capital, at the same time as it also satisfies an increase in GDP. Secondly: there is an intrinsic link between poverty eradication and the conservation of common natural resources, because the poor are the ones who benefit the most from the maintenance of natural capital. Finally, the transition to a green economy creates new jobs that, with time, would more than compensate the loss of jobs in the brown economy (the current model).

From an operational standpoint, the UNEP study shows that the amount needed to finance a global green economy is around 10% of total global investments per year. A reasonable allocation goal is $1.3 trillion per year, or 2% of global GDP, to promote a transition from brown to green economy. With this amount, the world can walk toward a decarbonisation of the energy matrix and a mitigation of the climate issue.

Economic policies and the institutional framework must be profoundly modified to show a signal of new times to economic agents.

We need new standards of behaviour and consumption if we want to leave our planet good children.

A RETURN TO THE CLASSICS

In its most primitive meaning, *oiko-nomia* concerned the duties and works (*nomos*) the head of the family should manage for the good of his house (*oikos*). It is interesting that economy and ecology are such closely related words. One is about managing the home and the other seeks to understand it in order to organize it.

The notion of home was extended to the planet. I return to Adam Smith in his book *The Theory of Moral Sentiments.*[22]

"The wise and virtuous man is at all times willing that his own private interest should be sacrificed to the public interest of his own particular order or society. He is at all times willing, too, that the interest of this order or society should be sacrificed to the greater interest of the state or sovereignty, of which it is only a subordinate part. [...] [He] must consider all the misfortunes which may befall himself, his friends, his society, or his country, as necessary for the prosperity of the universe, and therefore as what he ought, not only to submit to with resignation, but as what he himself, if he had known all the connexions and dependencies of things, ought sincerely and devoutly to have wished for."

Interestingly, Smith brings to mind a very broad and up-to-date concept. I particularly appreciate the remark: "he ought, not only to submit to with resignation, but as what he himself, if he had known all the connexions and dependencies of things, ought sincerely and devoutly to have wished for."

The economic world is not moving with the necessary urgency. There is already a critical mass of good thinkers reflecting on new paths to a future which would overcome the present crisis. A necessary condition for a new development cycle is for the return of the respectability of the currency.

The most serious matter in the current scenario, especially in developed countries whose currencies are said to be convertible (dollar and euro), is that because of the lack of adequate regulation of central banks and bodies monitoring macroeconomics, there was an inadvertent creation of virtual currencies. In some markets these "currencies," such as derivatives and others, flood the market and corrupt the foundations of a real economy.

We can see that the mechanisms for creating these "currencies" have been greatly facilitated by the manipulation of credits without an asset base in institutions themselves. These are the same institutions that issue and negotiate such bonds, either in the post-credit system or in the stock market roulette.

"Market benchmarks" are created to multiply the values of assets that, if one day are highly valued, the next day, without any connection to reality, decrease and produce losses for investors who often have no way of taking responsibility for it.

The conclusion is that these "currencies" which have no assets to guarantee them corrupt all reference values between the money supply and the real economy. A clear example can be seen in the creation of the junk bonds that caused the 2008 crisis, which still underlie all developed economies and those relying on the convertibility of their currencies.

The bubble of 2008 has not ended. We will certainly see other nuances of the same phenomenon in the coming years. However, we are heading toward a review of the references that created and sustained the global economy.

The fact is that the current economic model is unsustainable, but it will not end abruptly, or as T.S. Eliot tells us at the end of his poem

The Hollow Men: "This is the way the world ends / not with a bang but a whimper." That is, the changes will not happen explosively fast: they will happen through successive whimpers.

Now, what I advocate for is a change to a more desirable direction:

1. Currencies should be backed by real assets, consistent and subject to audit, like natural resources or in any set of items representing the real economy.
2. Financial institutions should be strictly regulated and conscious of their social role.

And finally, something which would take some time to be substantiated: pricing what we call natural capital so the value of renewable and non-renewable goods is reflected in exchange relations.

The key question is: why are we taking so long to act?

7

Possible paths

"The use of alternative and clean energy sources will be, in the twenty-first century, a conceptual change as important as the Copernican revolution in the sixteenth century, that is, the discovery of a new dimension to the planet."

In the lecture "What kind of world will the twenty-first century be?" May 9, 2002

This book was conceived in 2010 and 2011, driven by a sense of urgency about the decisive moment we are living through in this century. I sought to show the priorities currently presenting themselves: 1) The tools we develop for relations of exchange between nations, that is, the economic model; 2) The energy matrix necessary to produce, transport, and help man in all his activities; and 3) The manner in which nations organize themselves in order to maintain standards of freedom, ethics and social justice: the political model. These vectors need to be reformed and, therefore, must rely on the transformative power of our capacity for reflection.

Compassion for future generations should be more important than clinging to the comforts we presently have. By living in our microcosmos, we distance ourselves from our responsibility for the Earth. Man evolved from the flock to the tribe, and finally to States, but a nation is less important than the planet, which represents the highest level of political awareness we can desire.

We must expand our consciousness and move towards a society of "knowing," not of "having."

It will not be just with new technologies, sources of decarbonised energy and changes in the political model that we will be able to plan a new course for humanity.

The reinvention of the future requires multidisciplinary scientific improvements by adding to specific knowledge from other forms of knowledge which, when juxtaposed, could produce tools for a new world view.

This is the big question: a new world view that incorporates and reflects a new ethics, no longer theocentric or anthropocentric nor scientistic nor individualist. It should be integrative, planetary, considering humanity, all other species, and the Earth as a systemic unit.

The Earth is just one of billions and billions of planets that probably exist in the Universe, and it is precisely this apparent insignificance that gives us a unique identity. We are the Universe with a self-consciousness, we live on a rocky fragment marked by life in all its diversity. We are both special and insignificant to the cosmos of infinite size of which we can only have a glimpse. We and our home are one.

Evolution presupposes acquisitions and losses. We have developed the ability to learn from the past, question the present, and plan for the future. On the other hand, we have abandoned the possibility of life in the depths of the ocean, lost much of our sense of smell, hearing, and night vision... but we conquered reason! Man has been able to reach extreme territories like the moon and know parts of the Universe. We could even speculate, scientifically, that the end of the Universe in which we live will not be the end, because on the other side there will be the creation of a new Universe.

Either way, the emergence of a new consciousness must bring human responsibility for the part of creation that was given us. However, we are still divided and with individualistic, regionalist, and nationalist views. Now I wonder why we have not been able to synthesize a view of priorities for the future based on the the urgencies of the present? Could it be due to a lack of ethical perspective?

THE PATH OF THE COPS: AN ARCHAEOLOGY OF CLIMATE CONFERENCES

Since the Earth Summit, nations have been following the political and legal path of the great multilateral climate conferences in an attempt to contain the threat posed by climate change. The Climate Convention, adopted in Rio in 1992, is like a charter for guiding decisions on the subject. Since its entry into force in 1995, it has guided the institutional framework on the subject. Countries, or parties, who ratified it started to meet annually at the Conferences of the Parties, or COPs, to continue to shape the climate regime, either accompanying the implementation of decisions already taken, or by coming to new decisions.

After the Earth Summit, while the political difficulties of the process fed later conferences, emissions of greenhouse gases (GHG) grew year by year. The world economy practically ignored the alert contained in the first IPCC (*Intergovernmental Panel on Climate Change*) report of 1990, which drew attention to the seriousness of the problem and to the need to establish an agreement on reversing the increase in greenhouse gas emissions.

Annual CO_2 emissions in 1990 were 27.8 billion tons and the stock of atmospheric CO_2 reached 354ppm.[1]

The first Conference of the Parties took place only three years after the Earth Summit, in Berlin, when the signatories of the Climate Convention still had to ratify it internally – each one of the 154 countries, with their internal processes and idiosyncrasies. What we saw was that a naturally slow process became even slower; a slowness aggravated by the lack of political will – or by the presence of a political will *not* to come to any decision.

Annual CO_2 emissions in 1995: 28.9 billion tons.
Stock of atmospheric CO_2: 360ppm.

What the Berlin meeting left us was the establishment of a negotiating process extending for another two years, known as the Berlin Mandate, in order to set a limit on GHG emissions through a Protocol to be completed in 1997, in Kyoto.

A year after the COP-1 in Berlin, the IPCC launched another report, reinforcing the evidence of climate change and the need for strong political action to tackle the problem.

Annual CO_2 emissions in 1996: 29.3 billion tons. Stock of atmospheric CO_2: 362ppm.

GHG emissions grew while the bureaucracy got thicker and thicker.

At Kyoto-1997, after many changes in direction, a Protocol was finally reached (additional agreements subject to the Climate Convention) which established for the next decade a 5.2% reduction in relation to GHG emissions of 1990, but only for developed countries, technically called "Annex I Countries." The main instruments established in Kyoto were the carbon market, the CDM (Clean Development Mechanism) and instruments for their funding. In Kyoto, a legally binding instrument with targets was first established, but clearly a reduction of 5.2% for Annex I Countries would not suffice to revert an increase of global GHG emissions.

Annual CO_2 emissions in 1997: 29.7 billion tons.
Stock of atmospheric CO_2: 363ppm.

More ambitious numbers appeared at Kyoto: there was talk of up to 20% reduction in emissions from developed countries, but obviously this type of proposal had no political backing.

The truth is that the countries did not want to make commitments. The United States, the greatest emitter at the time, signed the agreement in 1997, but later President Bush publicly stated that he would not ratify the Kyoto Protocol, which further reduced the meagre goal of 5.2%.

The next COPs were marked by efforts to regulate the Kyoto Protocol. In practical terms, this meant establishing operational details for the rules, which were written in such a studiously vague manner as to please Greeks and Trojans. The conclusion of the regulations was scheduled for 2000 in The Hague.

Annual CO_2 emissions in 2000: 29.9 billion tons.
Stock of atmospheric CO_2: 369ppm.

There are moments in the COPs that are like jokes. The Hague was a total mess: it was impossible to finalize the regulations due to a total lack of consensus. After a very complicated episode, when Minister

Pronk, President of COP-6, pulled yet another independent proposal from his hat, the meeting was suspended and only resumed months later, in Bonn.

The so-called "COP-6 Reconvened" tried to reorganize the chaos, which postponed any conclusion until the next COP, the seventh, in Marrakesh in 2001.

The Marrakesh agreements finally concluded all regulations of the Kyoto Protocol. There was one highly controversial issue: deforestation was symptomatically left out of the discussions. Many interests were thrown into the same pot, including the issue of national sovereignty. What remained of extensive discussions was that only reforestation projects would be able to generate carbon credits.

Annual CO_2 emissions in 2001: 29.8 billion tons.
Stock of atmospheric CO_2: 371ppm.

It was only in 2005 that the Kyoto Protocol entered into force, with the entry of Russia to reach a minimum number of countries for the agreement to have any legal effect - just barely. The modest goal of 5.2% became even more useless with the absence of United States. *Faute de mieux*, everybody rushed for an embarrassed embrace.

The negotiation model of Kyoto is not a bad model, but as the United Stated did not ratify it, it has become a philanthropists club.

Annual CO_2 emissions in 2005: 32.9 billion tons.
Stock of atmospheric CO_2: 379ppm.

The first Kyoto commitment period began in 2008 and will end in 2012. The limitations of the Protocol have become more and more glaring. In addition to not including the United States, it did not include emissions from fast-growing developing countries like China and India, nor the emissions of GHG from international travel, nor other industrial gases with huge warming potential. In addition to the still limited utilization of financial mechanisms to promote transition to a low carbon economy, what we saw was an evolution of these rules, which became more and more limited to the initial formulation, increasingly distanced from evidence of climate change.

The increase in extreme events, like droughts, forest fires, floods, tornadoes and hurricanes corresponding to updated information about the worsening of climate change gave the Bali meeting in 2007 a critical importance not seen before since the Earth Summit in 1992.

A month before the Bali meeting, in collaboration with Rubens Ricupero, we warned in an article in the newspaper *O Globo*:

There are some fundamental questions about the Bali meeting. Will it be able to give impetus to the negotiations of more advanced and effective devices of the Kyoto Protocol? Will it be able to break the current deadlock in the struggle against the problems stemming from the energy matrix of fossil resources and the impacts of forest destruction and biodiversity? Facing the new situation, the Executive Secretary of the Convention on Climate Change, stated: "In order to avoid a gap after the end of the Kyoto Protocol's first phase in 2012, and the entry into force of the new institutional framework, the negotiations in Bali will need to conclude in 2009 to allow enough time for ratification." Will such a tight schedule be feasible?[2]

What remained from Bali-2007 was how the negotiations were divided into two tracks, the Climate Convention of 1992 and the Kyoto Protocol of 1997. The Bali Action Plan laid the groundwork for the legal form of themes that had been left out of the Protocol, such as reduction of emissions (or mitigation, in the climate jargon) of the United States and developing countries. The forest theme was retaken at another level with REDD+ (Reducing Emissions from Deforestation and Forest Degradation), which makes it possible to finance standing forests.

However, from Bali the arm wrestling between developing and developed countries also intensified. Developing countries put pressure on developed countries to assume substantial goals as a prerequisite to start a discussion about their own commitments. Developed countries only agreed to take up a more significant target if all the great emitters, including China, India and Brazil also took up their targets. Another deadlock: developed countries wanted to avoid discussions for a second period of the Kyoto Protocol, which was supposed to start in 2012 and intended to merge both tracks, the Convention's and the Protocol's, into a single Treaty which included all countries and all themes.

Annual CO_2 emissions in 2007: 34.1 billion tons.
Stock of atmospheric CO_2: 383ppm.

The negotiation models for climate change since 1992 have not changed because the bureaucratic apparatus which graciously accumulated over twenty years fed on numerous political deadlocks stimulated by interests, especially from the oil industries.

Over the last twenty years, the nuclei of interests widened, many of them veiled; many sectors have appropriated the theme. Today there is a gap between the expectations of civil society and the realities of the nominations within governmental structures and also within UN agencies. A large number of institutions began to swarm in this scenario without doctrine. There is a huge open system preventing progress in negotiations. Unfortunately, climate conferences have become more like tours and electoral platforms than a stage for decisions. There is a lack of the sense of urgency that the problem demands and an inability to reach an efficient technical-bureaucratic model.

The 2007 IPCC report showed the problem of climate change is even worse than scientists had estimated. We have very little time to stabilize the growth of GHG emissions.

The two real advances we have had in these twenty years were an increase in public awareness, and the intensification of the theme in all media. Those battles are won.

In Copenhagen-2009, public opinion saw an opportunity to exert pressure on heads of state. There was a good reason for that: it was hoped an agreement would be possible in relation to the first Kyoto Protocol period (2008-2012). All of us who were working on the theme knew this would not happen because conflicts of interest and power would prevent a possible agreement.

Annual CO_2 emissions in 2009: 34 billion tons.
Stock of atmospheric CO_2: 387ppm.

Perhaps we expected from Copenhagen an act of hope, but strong political statements did not happen. This was the reason why NGOs went there to put pressure on leaders. What we saw were proposals being emptied and, again, chaos. Heads of state traded on one side and delegates on the other, without coordination between them. And outside, there was increasing pressure from civil society and the press.

The COPs should be a forum for debates of ideas and life philosophies. The negotiation process, aimed at a consensus, could feed on an honest and fair debate – without pressure behind the scenes which would expose more clearly the interests and projects of each country. Western philosophical reason itself was born from the process of a public discussion of ideas, according to the thesis of Jean-Pierre Vernant, in *The Origins of Greek Thought*.[3] Discussions between warriors (*agon*) on the battlefield were later transferred to the public square (*agora*) in the new city-states (*polis*). Thus was born Politics, with a capital *P*, which is the art of reconciling the differences between citizens of equal dignity in favour of the common good. Now that politics and citizenship have reached a global level, it is time to return to the origins and refer to a time where all disagreements were exposed in public so that a common path could be built with them. The process of the Climate COPs will only end in concrete results if they are imbued with the spirit of the *agora* instead of remaining in the form of court intrigues, economic pressures, and domestic politicking. Nonconformity and will to action should be the predominant sentiments between delegates and heads of state.

One cannot repeat often enough how we must well understand that, when we talk about an increase in the average temperature on Earth, we are talking about the effects of temperature on climate. There is risk of a reverse ice age. In the remote geological past, the natural periods for species' adaptations to changes were much longer. Today those periods are much shorter. If we do nothing, we will be designing another world in which we do not know all the consequences for the lives of our descendants.

There is nothing substantial on the negotiating table from 2013 on. If there are no decisions to be implemented from 2013, we will have a dangerous lack of trading instruments. Countries are not interested in modifying their governance mechanisms.

My proposal goes towards building a new model of international trade that should make historical and recent polluters responsible in addition to encouraging clean energy sources and the pricing of natural capital.

Perhaps this new system can be administered by the World Trade Organization, an institution that operates in the real world of international trade.

Multilateral agreements with 154 countries on the basis of consensus are not viable as a leading solution for the problem of global emissions. Game theory explains the complexity of reaching an agreement between so many parties.

The result of the COPs must be considered a baseline, a minimum level of expectations while it is possible to reduce emissions through other bodies. The twenty greatest emitters are responsible for almost 80% of emissions. China and the United States today account for almost half of all annual emissions of greenhouse gases. They are the countries who should assume the leading role in reducing emissions and could bilaterally equate the issue.

Turning intentions into actions is a challenge to which I have committed myself in these twenty years after the Earth Summit. At the Rio conference, the private sector did not participate very much because it did not view the issue of climate change as their problem.

At Rio+20, the economic agents will probably be the strongest agents and actors. The private sector understands the risks, costs, and taxes; what it does not understand is ambiguity; proper signalling is important for the economic model to be reformed in a faster way. Surely this is one of the strong themes of Rio+20 — no more bureaucratic procrastination that does not solve anything.

REINVENTING THE WHEEL

The use of the wheel was the trigger for man to finally see his own responsibility for the planet and its resources, both natural and those produced by technologies. The wheel was created with the same impetus that allowed us to fly and see the Earth from another angle: globally.

We are a restless animal, not only in relation to our origin, but also in relation to our existence on this planet. Moving is part of our essence, and at the heart of the question about what we are doing here.

The Aristotelian view that nature had always been in perfect balance lasted until the late Middle Ages. The notion of anthropogenic interference in the natural realm came with the rational consciousness that characterized the Renaissance. In the sixteenth century in London, Gaunt conducted population censuses to determine the impact of population growth on the balance of the environment and the resources needed by the city.

Man's historical journey caused a gradual improvement in tools which, through evolution, structured the sciences which then, in their turn, produced technologies that once again were used to help us on our journey. The last two hundred years have sharpened our creative capacity and exponentially accelerated the advancement of science and technology.

However, the human capacity for creating solutions has been producing consequences contrary to their goals, and often we cannot control these effects. Generally, this happens out of arrogance because we do not consider the evolution of science and technology as open, continuous processes, and because we do not want to predict its perverse outcomes for the environment.

It is time for a paradigm shift because we have reached a new level of knowledge and experience. There is no determinism, no linearity – reality is experienced in a perpetual motion in which everything influences everything else.

THE BURIAL OF COPERNICUS

While I was writing this book I came across the following news: the remains of mathematician and astronomer Nicolaus Copernicus (1473-1543) were identified in a small Polish monastery and buried again on May 22, 2010, in the Cathedral of Frombork, 467 years after his death – this time with all the ceremonial and ecclesiastical honours.

Copernicus was the creator of the heliocentric theory, which considers the Sun as the centre of our planetary system. More than mathematical and astronomical studies, the disturbing thing in Copernicus's work was the philosophical change and conceptual change of man's relation to the Universe.

When Copernicus demonstrated the Earth is just one among several planets that orbit around the Sun, he caused much alarm. For if man was no longer the centre of the universe, the whole classical conception of the world would be put in check. How could a single individual, endowed only with reason and measurement instruments, contradict a tradition more than a thousand years old?

"We revolve around the Sun, like any other planet," Copernicus declared in 1514, but his book, *On the Revolutions of the Heavenly Spheres*, was published only in 1543. Copernicus refrained from disclosing his opinions, restricting the circulation of his theories to a select group of friends. Fear silenced his perspective. It fell to one of his former students to collaborate on the first edition of the book and submit it to a printer, thus starting the "Copernican Revolution."[4]

There is a tortuous episode in Copernicus's life; he was already physically and mentally weakened in 1543, the year of publication of his work. Stephen Hawking tells us: "The manuscript fell into the hands of Lutheran theologian Andreas Osiander. Osiander, hoping to appease advocates of the geocentric theory, made several alterations without Copernicus's knowledge and consent." He even changed the title at the top of the page, adding "hypothesis," and adding more of his own sentences to lessen the book's impact and dilute the final content of the work.

In the predominant world view of his time, man, created in the image and likeness of his Creator, should occupy a privileged place in the architecture of the world. There was a hierarchy in heaven that should be reflected in social hierarchies on earth. Copernicus's theory broke with this view and presented a major conceptual change, paving the way for modern Astronomy and affecting Science, Philosophy, and Religion.

Both Catholic and Protestant theologians attacked the Copernican theory, considering that a heliocentric universe contradicted the Bible. This idea would make people believe they were merely parts of a natural order, not the masters of nature, as in the traditional view. In the geocentric concept, arising from the Judeo-Christian world, man had the privilege of taming Earth according to his will and needs, because the planet was his inheritance in the order of creation.

"Like any other planet..." Copernicus's conclusion attacked the foundations of an entire world view, going beyond Theology and reaching social and individual consciousness. Copernicus's work was considered dangerous by the powers that be, but that did not prevent its appreciation by later scientists and philosophers, such as Francis Bacon, Johannes Kepler, Galileo Galilei, and Isaac Newton.

In the sixteenth century, the philosopher and Dominican friar Giordano Bruno was an enthusiastic advocate of the Copernican theory, going beyond it by speculating that there might be other worlds with intelligent life in the universe. It was too much for the *status quo* of the time! Giordano Bruno ended up being burned at the stake by the Inquisition.

The discoveries of Copernicus formed the basis for the Scientific Revolution which, in turn, paved the way for the industrial and technological revolution, shaping the world we live in today.

The Catholic Church only officially accepted the Copernican theory in 1922, a period when Science had already surpassed classical physics and evolved to another paradigm with Einstein, unravelling the secrets of matter, light, and space-time.

The Copernican hypothesis was proved true in the years following his death, but the philosophical implications that we are not masters, but small parts of a much larger whole, still seems to not be fully assimilated to this day. Stephen Hawking, in the book *On the Shoulders of Giants*, mentions the German scientist Johann Wolfgang von Goethe, who was perhaps the most eloquent of all on writing about Copernicus's contributions:

"The world had scarcely become known as round and complete in itself when it was asked to waive the tremendous privilege of being the centre of the universe. Never, perhaps, was a greater demand made on mankind — for by this admission so many things vanished in mist and smoke! [...] No wonder his contemporaries did not wish to let all this go and offered every possible resistance to a doctrine which in its converts authorized and demanded a freedom of view and greatness of thought so far unknown, indeed not even dreamed of."[5]

Copernicus's brilliance was very belatedly recognized by the Church, only after four centuries. However, if the world takes four hundred years to recognize and attack the causes of climate change and its effects on Earth, our species will no longer even be here.

Denying scientific truths is dangerous. If there are no reforms in our way of life, it will not take more than forty years for us to have a completely different planet. This transformation will happen with a huge impact on all forms of life. We do not know what will become of humanity, but the Earth will continue to exist and revolve around the Sun.

In the future, will the second burial of Copernicus be considered a symbolic moment for a world that experienced a new twist of awareness about man's role in the universe? I hope so.

THE REDISCOVERY OF ETHICS

The transfer of knowledge from the past to the future does not happen automatically; it happens in the transmission and modification of culture, of all human actions that pertain to everyday life. Historically, knowledge has suffered advances and setbacks, suppression and stimulation, according to its ability to provide tools for the dominant values of society. The ancient Babylonians invented Mathematics and Astronomy, but only millennia later would calculations serve as the basis for the creation of advanced technologies. The Greeks invented democracy and public participation about 2500 years ago, but only in the mid-twentieth century were these political principles established.

The civilization based on anthropocentric development principles is exhausted. The economy, shaped by the philosophical, religious, and moral principles generated in the seventeenth and eighteenth centuries, put man as beginning and end, justifying the continued use of wealth to produce more wealth, ignoring the responsibility to creation.

The recognition of a creative force in the universe gives man the desire to transform and celebrate his bond to society and the cosmos. Our individual organism is deeply connected with the whole universe.

The human metabolism produces compassion: a mother feels compassion for her child, and society should feel compassion for those who need support. We find that, in many situations, individual interests have supplanted collective interests.

Ethics is the link in the consolidation of a project that combines technological advances with the transformation of civilization itself.

This ethics will be much more complex than the ethics of Plato and Aristotle; it will consolidate technological advances into common purposes, in which the trade-offs of military budgets and consumerism will be exchanged for the direct responsibility of each individual with respect to their choices.

We must differentiate ethical issues from moral issues. Morality is more related to contingencies, with the particular, while ethics is part of universal thought. Morality changes with time and customs, but the basic principles of ethics remain.

I have sought to show the fundamental dilemmas of the current crisis and proposals for possible ways out. What will happen in 2050 depends on the actions we take in the current decade.

I shall close with an imagination exercise on what a journey into the future could be like, considering the worst and the best case scenarios.

To write the last section of this book I will refer to a series of conferences promoted by the Eva Klabin Rapaport Foundation in 2001, called "Comprehending the twenty-first century." We were then in the first year of a new century, and a combination of concern and hope crossed our hearts. The participants in the discussion were (Eva died in 1991) the then President of the Foundation, Helio Jaguaribe, the physicists Prof. Mario Novello and Prof. Luiz Alberto Oliveira, the biochemist Prof. Leopoldo de Meis and myself, who gave a conference I called "Earth: the Limit of Sustainability." While the other panelists exposed subjects as complex and fascinating as the creation of the universe, the relationship of man with technology, and the return of religious fundamentalism, I sought to get across a message that, in retrospect, can be seen as the embryo of this book. At the end of my presentation about the environmental challenge for the twenty-first century, I imagined possible scenarios for the year 2050. I will cross a stream of time backed up by scientific projections.

THE WORST CASE SCENARIO

We are in the first hour of the year 2050. While the last bells announcing the end of 2049 ring in New York to welcome the New Year, people pause to reflect on how things have changed during their lives. With the global temperature 3°C higher than at the turn of the twentieth to the twenty-first century, the world is now a very different place. Electronic archives keep the memory of the former world alive.

Tuvalu and Kiribati flooded

Millions of people have fled from the low atoll areas of Tuvalu, Kiribati, and the Marshall Islands due to an increase of 80 cm in sea level. The Marshall Islands were the first to disappear, and their first inhabitants now live in poverty, in Indonesia. There are huge political conflicts caused by the new climate diasporas – only the residents of Kiribati were allowed to move to New Zealand.

The Museum of Modern Art in New York has just set up an exhibition called "Extremes", already drawing a record audience, showing the great environmental disasters in the last decades. In Europe, parents tell their children about a time when the Alps were covered in snow and had ski slopes. The Mont Blanc really was a white mountain. Only the largest glaciers remain. In Asia, the Himalayas have lost nearly a third of their ice, and last year the Ganges River became dry for the first time. The populations of India and Bangladesh panicked and, in the largest migration in human history, almost 300 million people are currently moving toward Europe.

In this half of the twenty-first century, big cities like Rio de Janeiro, São Paulo, Paris and Hong Kong, are being increasingly degraded as basic services become scarce.

Climate change officially out of control

Scientists announced at the end of last year that climate change is out of control, officially without limit. The Arctic started to melt fifty years ago and is currently releasing ten times more carbon into the atmosphere than all anthropogenic emissions. The presidents of the United States and China made an emergency meeting on New Year's Eve, but since the latest round of conversations on the climate ended in acrimony, they were unable to agree on any new measure. The unfolding of the COP-46 meeting was even more embarrassing for the U.S. delegation because the meeting was held in the Japanese city of Kyoto, where an agreement signed half a century ago was not able to reach concrete goals.

China and India now consume more than the United States, whose economy has been shaken by recent statements from scientists and is still recovering from the almost total failure of the insurance industry, after the destruction of Washington D.C. by the Exxon Hurricane. The shock was enough to cancel the programme of U.S. aid to Bangladesh, where five million people died in the storms of October 2042.

Bombardment of electric plants narrowly averted

Last October, a potential war was averted when the United States withdrew the threat of bombing Shanghai unless China closed 200 of its coal-fired plants. China still insists that it has a moral right to pollution, because it was the U.S. that originally caused the problem. Therefore, Americans should pay for the carbon emitted by the Chinese. Millions were removed from the cities of Wuhan and Chongqing when the world's largest river, the Yangtze, dried up in 2036, causing the collapse of the Three Gorges hydroelectric dam. Today, the major causes of international law are governed by the nations affected by climate change against the major carbon emitters.

Polar bears have been reconstructed at the Museum of Natural History in New York, alongside the dinosaurs. The last bear was seen in northern Canada at the end of the 2020s. The poles became warmer faster than the rest of the planet, and therefore the increase in temperature in those regions is already at six degrees. The entire Arctic Ocean is free of ice during the summer.

Chinstrap and Emperor penguins have disappeared from Antarctica, but these extinctions are nothing compared to what is happening in the Amazon. More than three quarters of what was once the largest rain forest were destroyed by fire, and amid the ashes there is only desert. In 2011, a study funded by the World Bank and published in the journal *BioScience* already warned that there will be a reversal point when 20% of the Amazon is hit by deforestation, leading to a state of desertification called Amazon Dieback.

The end of the forests

Zoologists estimate that ten species of monkeys, 200 kinds of birds and many tree species are already gone forever. The end of the forests has also had greater effect on climate change, with more carbon entering the atmosphere over the last decade than in the whole twentieth century. This increased the atmospheric carbon dioxide level to 964ppm, three times more than its level during the pre-industrial era. The ancient permafrost of Siberia also melted, releasing immense amounts of methane in the air, intensifying the greenhouse effect.

A new global sect has been dominating the religious landscape and proclaimed Gaia its only goddess. The "Gaians" encourage terrorist attacks, particularly on industrial plants, oil rigs, and coal plants. They proclaim a return to what they call the Earth Age and the abandoning of technological apparatus in favour of a rustic life, living in harmony with the animals and plants which remain. Malthusian doctrine re-emerges: the radicals propose that instead of 10 billion people, the population should regress to 1 billion, a figure they consider feasible for a healthy relationship with the planet.

And who will choose those who will remain?

THE BEST CASE SCENARIO

It is New Year's Eve for 2050, which is being celebrated much more joyfully than usual over the whole world. Actually, this is the biggest party since the turn of the millennium in 2000, and by no numerical accident, because 2050 was the deadline given by the United Nations for the global economy to dramatically reduce the burning of fossil fuels. And to the surprise of all, especially the cynical, this goal was met. Last year the average global temperature was the highest in history, two degrees above the levels of the late twentieth century, but this record will likely not be hit again.

China and the United States, the largest emitters of greenhouse gases, sat together at the negotiating table, permitting a real advance in the signing of the International Treaty on Climate Change of 2030, negotiated after the old Kyoto process of 1997. They reached a pact between rich and poor countries, in which all recognized that they had the same right to the atmosphere with equal and not differentiated responsibilities.

In 2042, at the Rio+50 Conference, there were great festivities in Rio de Janeiro to celebrate the fiftieth anniversary of the Earth Summit, the Rio Conference, an event considered by historians as the precursor of the green revolution.

Ethiopia, Somalia, and Tajikistan uninhabited because of drought

The signing of the Treaty is still remembered as a moment of great emotion in history. When all world leaders signed the document, the delegation of the countries of the flooded islands, Tuvalu and Kiribati, held a ceremony in which all countries had to be involved. The representatives for Ethiopia, Somalia, Afghanistan, and Tajikistan joined them because their countries were also no longer considered habitable because of persistent drought.

Historical treaty wins the day

Under the terms of the Treaty, the populations of countries declared uninhabitable by the United Nations received assistance in Europe and in Central and South America, now the superpowers of biodiversity. Decades of severe immigration policies – whose goal was to keep environmental refugees out from rich countries – have been changed.

To facilitate the International Treaty on Climate Change of 2030, the United States and European countries have pledged to tax carbon and reallocate military budgets to projects for solar and wind energy. The *monnaie de compte* is now based on a basket of natural capital.

Human well-being as reference

The cities of countries which have become developed, like Brazil, China, and India, are reference points for welfare and civilization. Underground networks of garbage collection go through their main cities, while hydrogen-powered trains are efficient means of transport.

In all Brazilian cities, energetically self-sufficient popular green neighbourhoods were built, with wind and solar generators. Bike lanes took the place of grand boulevards. The current global celebrities are three scientists who have just shared the Nobel Environment Prize: a Brazilian, a Chinese, and an Indian, the inventors of the solar-wind generators that sparked the environmental revolution.

Three decades ago, when the deadline was set for reducing the use of fossil fuels, economists thought that it would bankrupt the world. Part of the problem was the numbers – until the Gross Domestic Product Index was replaced by the Human Well-being Index, it was impossible to differentiate good growth from poor growth. Discussions on a new index triggered a fierce fight in the World Economic Forum in 2024. Fortunately, a consensus was reached.

According to the former indicator, known as GDP, even oil spills and pollution appeared as growth, and all human needs were subordinated to the pursuit of corporate profits. But, like quality of life, food, shelter, education, and other indicators of well-being became referential, people started to think about development in a different way. Market forces soon led to the use of clean technologies. Even Saudi Arabia, a former oil producer, was able to employ many former employees from oil platforms, now obsolete, in solar energy plants installed throughout the deserts.

Too late for the Pygmies

But the International Treaty on Climate Change came too late for the inhabitants of tropical forests, such as the pygmies of the Congo and the Yanomami of Brazil. Both groups were important in the mobilization of support for the signing of the Treaty, and also in obtaining an international prohibition against the felling of ancient forests in 2030. But the lack of rain caused many forests to continue burning throughout the dry seasons, leading nearly to the extinction of many game animals on which they depended.

Other sad news came from the Arctic, where global warming resulted in a temperature increase of three degrees. The last wild polar bear was seen near Point Lay in Alaska, almost two decades ago – the only surviving bears are in zoos throughout the world. It is expected that when temperatures are stabilized, they can be re-introduced to their former homes.

Not 100% yet

Sea levels will continue to rise in the next thousand years because the absorption of heat by the oceans is much slower than the atmosphere. But the icing on the West Antarctic is still stable, and Greenland melts at a slower speed than science had previously indicated.

We still do not know what will happen for several decades. But what is really creating this festive atmosphere for New Year's Eve 2050 is the fact that humanity was able to unite against the greatest challenge in history.

Jewish, Christian, Muslim, Buddhist, Hindu, Sikh, and Taoist leaders meet annually in a large festival that celebrates religious diversity and gives thanks for the natural resources the Earth generously offers.

On January 1, 2050, New Year's Eve was not the highlight on the main digital newspapers, but the news that for the first time in three centuries, the level of CO_2 in the atmosphere had significantly decreased. The target set at the International Treaty on Climate Change was reached. But only just.

References

ABRANCHES, Sérgio. *Copenhague Antes e Depois*. Rio de Janeiro: Civilização Brasileira, 2010.

ALQUÉRES, José Luiz (org.). *Energia Para Gerações*. Rio de Janeiro: Shell Brasil Ltda., 2003.

ANDRADE, Carlos Drummond de. *Uma Pedra no Meio do Caminho. Biografia de um Poema*. São Paulo: Instituto Moreira Salles, 2010.

ARNT, Ricardo (org.). *O Que os Economistas Pensam sobre Sustentabilidade*. São Paulo: Editora 34, 2010.

ASSOULINE, Pierre. *Rosebud Fragmentos de Biografias*. Rio de Janeiro: Rocco, 2010.

AZEVEDO, Francisco Ferreira dos Santos. *Dicionário Analógico da Língua Portuguesa Ideias Afins/Thesaurus*. Rio de Janeiro: Lexicon, 2010.

Banco Nacional de Desenvolvimento Econômico e Social: *Amazônia em debate: oportunidades, desafios e soluções*. Rio de Janeiro: BNDES, 2010.

BECHARA, Evanildo. *Minidicionário da Língua Portuguesa atualizado pelo Novo Acordo Ortográfico*. Rio de Janeiro: Editora Nova Fronteira, 2009.

BECKER, Bertha. STENDER, Claudio. *Um futuro para a Amazônia*. São Paulo: Oficina de Textos, 2008. (Série Inventando o Futuro)

BERGREEN, Lawrence. *Marco Polo, de Veneza a Xanadu*. Rio de Janeiro: Editora Objetiva, 2007.

BERLIN, Isaiah. *Estudos Sobre a Humanidade. Uma Antologia de Ensaios [The Proper Study of Mankind: An Anthology of Essays]*. Rio de Janeiro: Cia. Das Letras, 2002.

Bíblia Hebraica . São Paulo: Editora e Livraria Sêfer Ltda, 2006.

BRITO, Sergio. *Desafio amazônico: o futuro da civilização dos trópicos.* Brasília: UnB/CNPq, 1990.

BROWN, Lester. *Plano B 4.0. Mobilização para salvar a civilização [Plan B 4.0 – Mobilizing to Save Civilization].* São Paulo: New Content, 2009.

BROWN, Cynthia Stokes. *A Grande História do Big Bang aos Dias de Hoje.* Rio de Janeiro: Civilização Brasileira, 2010.

CALVINO, Italo. *As Cidades Invisíveis.* São Paulo: Companhia das Letras, 2000.

CALVINO, Italo. *Invisible Cities.* Orlando: Harcourt Trade Publishers, 1974.

CAMPOS, Roberto. "Bretton Woods, FMI, BIRD, Havana e GATT: A Procura da Ordem Econômica do Após-Guerra". Brasília: Boletim de Diplomacia Econômica – MRE, February 19, 1995.

CARDOSO, Fernando Henrique. "Global Governance in the XXI Century". Series of debates about the international political and economic outlook. London: London School of Economics and Political Science (LSE), November 16, 2004.

CARDOSO, Fernando Henrique. "The need for Global Democratic Governance: a perspective from Latin America". Washington, DC: Library of Congress, February 22, 2005. "Henry Kissinger Lectures in Foreign Policy and International Relations".

CHANDLER, Arthur. "Fanfare for the New Empire. The Paris Exposition of 1855". World's Fair magazine, vol. VI, nº 2, 1986.

Comission for Biodiversity, Ecosystems Finance and Development: "Latin America and the Caribbean. A Biodiversity Superpower". New York: United Nations Development Programme, 2010.

United Nations Conference on Environment and Development: *Convention on Biological Diversity.* Rio de Janeiro: United Nations, 1992. http://www.onu-brasil.org.br/doc_cdb1.php and http://www.cbd.int/convention/text/

COOK, Earl. "The flow of energy in an industrial society", in *Man, Energy and Society.* W. H. Freeman Edition, 1976.

DIAMOND, Jared. *Colapso. Como as sociedades escolhem o fracasso ou o sucesso.* Rio de Janeiro: Record, 2005.

DIAMOND, Jared. *Collapse. How societies choose to fail or succeed.* New York: Viking, 2005.

DONELLA H. Meadows *et al. The Limits to Growth. A report for the Club of Rome's Project on the Predicament of Mankind*. Washington, DC: Potomac Associates, 1972.

DREXHAGE, John. MURPHY, Deborah. *Sustainable Development: from Brundtland to Rio 2012*. New York: UN Headquarters, September 2010.

DURANT, Will. *Nossa Herança Oriental*. Rio de Janeiro: Editora Record, 1963.

ELIOT, T.S. *Four Quartets*. New York: Harcourt Brace Jovanovich, 1943.

ELIOT, T.S. *Obra Completa Volume I. Poesia*. São Paulo: Arx, 2004. (tradução de Ivan Junqueira).

ELIOT, T.S. *Poesia*. Rio de Janeiro: Nova Fronteira, 2001. (Tradução de Ivan Junqueira.)

ELKINGTON, John. "Enter the Triple Bottom Line". IN Henriques, Adrian. Richardson, Julie: *The Triple Bottom Line, Does it All Add Up? Assessing the Sustainability of Business and CSR*. London: Earthscan, 2004.

ELKINGTON, John. *Cannibals with Forks. The Triple Bottom Line of 21st Century Business*. Oxford: Capstone Publishing, 1997.

ERNEST, Renan. *Qu'est-ce qu'une nation?* Conference at the Sorbonne, March 11, 1882. (p. 10-11)

ERNEST, Renan. "What is a Nation?" IN Eley, Geoff and Suny, Ronald Grigor, ed. 1996. *Becoming National: A Reader*. New York and Oxford: Oxford University Press, 1996.

FRIEDLINGSTEIN et al. *Update on CO_2 Emissions*. Nature Geoscience, doi:10.1038/ngeo_1022. Published on-line, November 21, 2010. Global Carbon Project: *Carbon Budget and Trends 2009*. www.globalcarbonproject.org/carbonbudget

Fundação Brasileira para o Desenvolvimento Sustentável [Brazilian Foundation for Sustainable Development] (FBDS): "Estimativas da oferta de recursos hídricos no Brasil em cenários futuros de clima (2015-2100)." Commented by Eneas Salati (editor), Walfredo Schindler, Daniel de Castro Victoria, Eneida Salati, João Carlos Simanke de Souza and Nilson Augusto Villa Nova.

GALLAI, N. *et al. Economic Valuation of the Vulnerability of World Agriculture Confronted with Pollinator Decline*. Ecological Economics (2009), Vol. 68 (3): 810-21.

GIANETTI, Eduardo. *O Livro das Citações*. São Paulo: Companhia das Letras, 2008.

GIDDENS, Anthony. *A Política da Mudança Climática*. Rio de Janeiro: Jorge Zahar Editora, 2010.

GLEISER, Marcelo. *Criação Imperfeita. Cosmo, Vida e o Código Oculto da Natureza*. Rio de Janeiro: Record, 2010.

HALÉVY, Daniel. *Essai sur l'accélération de l'histoire*. Paris: Editions de Fallois, 2001.

HAWKING, Stephen. *On the Shoulders of Giants: The Great Works of Physics and Astronomy*. Philadelphia: Running Press, 2003.

HAWKING, Stephen. *Os Gênios da Ciência. Sobre os Ombros de Gigantes. As mais importantes ideias e descobertas da física e da astronomia organizadas e comentadas pelo mais famoso físico da atualidade*. Rio de Janeiro: Editora Campus-Elsevier, 2005.

Hebrew Bible. A Hebrew-English Bible According to the Masoretic Text and the JPS 1917 Edition. 2005 http://www.mechon-mamre.org/p/pt/pt0.htm

HOBSBAWM, Eric. *Age of Extremes*, London: Abacus, 1995. (p. 49).

HOBSBAWM, Eric. *Bandidos*. São Paulo: Paz e Terra, 2010.

HOBSBAWM, Eric. *Era dos Extremos*, São Paulo: Companhia das Letras, 1995.

Instituto Antônio Houaiss. *Mini Houaiss Dicionário da Língua Portuguesa*. Rio de Janeiro: Objetiva, 2003.

International Energy Agency/OECD: "Clean energy progress report". IEA/OECD, 2011.

ISAACSON, Walter. *Kissinger: A Biography*. New York: Simon & Schuster, 1992.

JÖHR, Hans. *O Verde é o Negócio*. São Paulo: Saraiva, 1994.

JUDT, Tony. *O Mal Ronda a Terra. Um Tratado Sobre as Insatisfações do Presente [Ill Fares the Land]*. Rio de Janeiro: Editora Objetiva Ltda, 2010.

KANT, Emmanuel. *Ideia de uma História Geral com um Propósito Cosmopolita*. Covilhã: LusoSofia Press, 2011.

KENNEDY, Robert. *Remarks at the University of Kansas*. Kansas: March, 1968.

KEYNES, John Maynard. "National Self-Sufficiency," *The Yale Review*, Vol. 22, nº 4 (June 1933).

KISSINGER, Henry. *Diplomacy*. Norwalk: The Easton Press, 1994.

KLABIN, Israel.SALATI, Eneas. "Algumas Observações Sobre a Avaliação dos Impactos Ambientais e Hidrológicos do Projeto da Hidrovia Paraguai-Paraná" In *O Projeto de Navegação da Hidrovia Paraguai-Paraná. Relatório de uma Análise Independente.* Brasília: EDF/CEBRAC, July 1997. Israel Klabin's collection.

LAGO, André Aranha Corrêa do. *Estocolmo, Rio, Joanesburgo. O Brasil e as Três Conferências Ambientais das Nações Unidas.* Brasília: Instituto Rio Branco and Fundação Alexandre de Gusmão, 2007.

LANDES, David. *Prometeu Desacorrentado. Transformação Tecnológica e desenvolvimento industrial na Europa Ocidental, desde 1750 até a nossa época.* Rio de Janeiro: Nova Fronteira, 1994.

LINS, Clarissa. ZYLBERSZTAJN, David (org.). *Sustentabilidade e Geração de Valor. A Transição para o Século XXI.* Rio de Janeiro: Campus-Elsevier, 2010.

MANDER, Jerry.GOLDSMITH, Edward (eds.). *The Case Against the Global Economy and for a turn toward the local.* San Francisco: Sierra Club Books, 1996.

MARCOVITH, Jacques (coordenador) *et al. Economia da Mudança do Clima no Brasil: Custos e Oportunidades.* São Paulo: IBEP Gráfica, 2010.

MARCOVITCH, Jacques. *A Gestão da Amazônia. Ações Empresariais, Políticas Públicas, Estudos e Propostas.* São Paulo: Editora da Universidade de São Paulo, 2011.

MARINETTI, Filippo. *Manifesto of Futurism.* Paris: *Le Figaro,* February 20, 1909. (Translated by James Joll).

MAY, Peter (org.). *Economia do Meio: Ambiente Teoria e Prática.* Rio de Janeiro: Campus-Elsevier, 2010.

MYERS, Norman. *Ultimate Security: the Environmental Basis of Political Stability.* New York: W.W. Norton & Co., 1993.

OLIVEIRA, Lúcia Lippi (org.). *Cidade: História e Desafios.* Rio de Janeiro: Editora FGV, 2002.

PÁDUA, José Augusto. *Um Sopro de Destruição: Pensamento Político e Crítica Ambiental no Brasil Escravista.* Rio de Janeiro: Jorge Zahar Editora, 2002.

PERLMAN, Janice. *Favela. Four Decades Living on the Edge in Rio de Janeiro.* New York: Oxford University Press, 2010.

PETIT, Karl. *Le Dictionnaire des Citations du Monde Entier.* Verviers (Belgique): Editions Gerard & Co,s.d. Collection Marabout, 1978.

POLO, Marco. *The travels of Marco Polo, the Venetian: the translation of Marsden revised, with a selection of his notes*. London: H. Bohn, 1854. (p. 229-230)

Population Division, UN-DESA. *2010 Revision of World Population Prospects Press Release*. New York, UN Department of Economic and Social Affairs, May 2011.

Prefeitura do Rio: *Leis e Decretos. Administração Klabin*. Rio de Janeiro: Secretaria Municipal de Administração/Superintendência de Documentação, 1980.

PRINS, Gwyn *et al*. *The Hartwell Paper. A new direction for climate policy after the crash of 2009*. London: Institute for Science, Innovation and Society University of Oxford/London School of Economics and Political Science, 2010.

PUTNAM, Robert. *Bowling Alone: America's Declining Social Capital*. Journal of Democracy 6:1, Jan, 65-78, 1995.

SACHS, Jeffrey. *A Riqueza de Todos. A construção de uma economia sustentável em um planeta superpovoado, poluído e pobre*. Rio de Janeiro: Nova Fronteira, 2008.

SCHINDLER, Walfredo (org.). 1º *Seminário Brasileiro sobre CCS – Carbon Dioxide Capture and Storage. Um debate sobre os desafios dessa nova tecnologia. Rio de Janeiro:* Fundação Brasileira para o Desenvolvimento Sustentável – FBDS/Shell Brasil, 2009.

SCHMIDT, Augusto Frederico. *Poesia Completa. 1928-1965*. Rio de Janeiro: Topbooks/Faculdade da Cidade, 1995.

SMITH, Adam. "The Theory of Moral Sentiments". London: A. Miller, 1790, 6th edition. (VI.II.46) http://www.econlib.org/library/Smith/smMS.html

SPALDING M.D. *et al*. "World Atlas of Coral Reefs". Berkeley: University of California Press, 2001. UNEP – Coral Atlas. UNEP – WCMC Coral Reef Unit.

STERN, Nicholas. *O Caminho Para um Mundo Mais Sustentável*. Rio de Janeiro: Campus-Elsevier, 2010.

STERN, Nicholas. "A conta ficou mais alta" – interview to Ana Luiza Herzog. *EXAME*. São Paulo, August 11, 2010.

STRONG, Maurice. *Where on Earth are We Going?* Toronto: Random House of Canada, 2000.

TEEB for National and International Policy Makers. Summary: Responding to the Value of Nature. TEEB – The Economics of Ecosystems and Biodiversity, 2009.

The Earth Charter Initiative: *Carta da Terra*. Paris: UNESCO, 2000. http://www.cartadaterrabrasil.org/prt/text.html and http://www.earthcharterinaction.org/content/pages/Read-the-Charter.html

The Worldwatch Institute: *2010 State of the World. Transforming Cultures From Consumerism to Sustainability*. New York: W.W. Norton & Company Inc, 2010.

TOYNBEE, Arnold. *Um Estudo da História [A Study of History]*. Rio de Janeiro: Martins Fontes, 1986.

TUCKER, Mary Evelyn.GRIM, John A. *Daedalus: Journal of the American Academy of Arts and Sciences. Fall 2001. Religion and Ecology: Can The Climate Change?* Cambridge: American Academy of Arts and Sciences, 2001. Issued as Volume 130, Number 4 of the *Proceedings of the American Academy of Arts and Sciences.*

UNEP, 2011, *Towards a Green Economy: Pathways to Sustainable Development and Poverty Eradication – A Synthesis for Policy Makers*. www.unep.org/greeneconomy

United Nations Department of Economic and Social Affairs Population Division: *World Population to 2300*. New York: United Nations, 2004.

U.S. Census Bureau: "U.S. and World Population Clocks". Washington: U.S. Census Bureau, 2011. www.census.gov/main/www/popclock.html

U.S. Energy Information Administration: "International Energy Outlook 2010". Washington DC: US Department of Energy, 2010.

VERNANT, Jean-Pierre. *As Origens do Pensamento Grego*. Rio de Janeiro: Bertrand Brasil, 2002.

VERNANT, Jean-Pierre. *The Origins of Greek Thought*. Ithaca: Cornell University Press, 1982.

WALKER, Brian.SALT, David. *Resilience Thinking. Sustaining Ecosystems and People in a Changing World*. Washington: Island Press, 2006.

World Bank Poverty Reduction and Equity group: *Food Price Watch*. World Bank: April 2011.

World Bank: *Governance and Development*. Washington, DC: World Bank, 1992.

World Wildlife Fund: *Living Planet Report 2010. Biodiversity, biocapacity and development*. Gland (Switzerland): WWF, 2010.

ISRAEL KLABIN'S COLLECTION

KLABIN, Israel et al. "O Futuro da Amazônia é o nosso futuro". *Folha de S.Paulo*, São Paulo, 2005.

KLABIN, Israel. "O Novo Nome Para a Paz". *O Globo*, October, 2004.

KLABIN, Israel. "Que tipo de mundo será o século XXI?" Lecture given at AMATRA, May 9, 2002.

KLABIN, Israel. *Terra: Limites da Sustentabilidade*. Rio de Janeiro: Eva Klabin Rapaport Foundation, 2001. (Lecture given on November 29).

KLABIN, Israel. "Remarks". OCDE: Paris, 25 de janeiro de 2001. *International Workshop on Market Creation for Biodiversity Products and Services.*

KLABIN, Israel, lecture: *FAO Advisory Committee on Paper and Wood Products, Fortieth session.* São Paulo: FAO, April 28, 1999.

KLABIN, Israel. "Compreendendo a Cultura no Desenvolvimento Sustentável: investindo em legados culturais e naturais", World Bank and UNESCO, September 28-29, 1998, World Bank Annual Meeting.

CARDOSO, Fernando Henrique. "Carta ao Senhor Israel Klabin". Brasília, October 23, 1997.

KLABIN, Israel. "Carta a Fernando Henrique Cardoso". Rio de Janeiro, September 17, 1997.

KLABIN, Israel. "Marte e a destruição da Amazônia". *O Globo*, Rio de Janeiro, July 10, 1997.

KLABIN, Israel. "Energia renovável e eficiência energética para o desenvolvimento sustentável". Palestra Seminário REIA, July 9, 1997.

KLABIN, Israel. "Uma agenda para florestas tropicais". *Folha de S.Paulo*, São Paulo, June 26, 1997.

KLABIN, Israel. "A controvertida hidrovia Paraguai-Paraná". *Folha de S.Paulo*, São Paulo, March 14, 1997.

KLABIN, Israel. "Financiamento para Projetos de Meio Ambiente – dificuldades e oportunidades". Workshop Meio Ambiente – Oportunidades de Negócios e Riscos a Curto e Longo Prazo – Nov 12, 1996.

KLABIN, Israel. "Desenvolvimento Sustentável – Harmonia Entre Ecologia e Economia." Fórum da Sociedade Civil para o Meio Ambiente ACRJ/Light 1994 – Painel sobre Cultura e Meio Ambiente Relações com o Desenvolvimento Econômico.

KLABIN, Israel. "Preservando o Futuro". *Jornal do Brasil*, Rio de Janeiro, June 1, 1992.
KLABIN, Israel. "A sociedade civil e o setor privado no desenvolvimento sustentável", Rio de Janeiro: SID – Sociedade para o Desenvolvimento Internacional, May 27, 1992. (Marina Palace Hotel).
KLABIN, Israel. "Inaugural speech." Official opening ceremony of Rio+5 Forum, 1997.
KLABIN, Israel. "State, Peace and Power: an exercise on democracy (the three basic rights)," Bombay, India, 1990.
GUDIN, Eugênio. "Israel Klabin no Leme do Burgo Podre". *O Globo*, Rio de Janeiro, 1979.
KLABIN, Israel. "Speech," September 27, 1979.
Publication by the City of Rio de Janeiro/Secretaria Municipal de Planejamento e Coordenação Geral.
KLABIN, Israel. "The Liberal Issues and The Third World". Rio de Janeiro, September, 1966.

INTERVIEWS

Conducted by Cristina Aragão and Leonardo O'Reilly Brandão with the author and the FBDS team – Brazilian Foundation for Sustainable Development:
Walfredo Schindler – director superintendent.
Clarissa Lins – executive director.
Agenor Mundim – project coordinator in the area of energy.
Eneas Salati – technical director from 1992 to 2011.
Branca Americano – special advisor to the President of FBDS.

GOVERNMENTAL, NON GOVERNMENTAL AND SCIENTIFIC INSTITUTIONS

FAO/UN – Food and Agriculture Organization of the United Nations
FBDS – Fundação Brasileira para o Desenvolvimento Sustentável [Brazilian Foundation for Sustainable Development]
Footprint Network
Fundação Eva Klabin Rapaport
Global Carbon Project

IBGE – Instituto Brasileiro de Geografia e Estatística [Brazilian Institute of Geography and Statistics]
IEA/OECD – International Energy Agency
INPE/MCT – Instituto Nacional de Pesquisas Espaciais do Ministério da Ciência e Tecnologia [National Institute for Space Research of the Brazilian Ministry of Science and Technology]
IPAM – Instituto de Pesquisa Ambiental da Amazônia [Amazon Environmental Research Institute]
NASA – National Aeronautics and Space Administration
NOAA – National Oceanic and Atmospheric Administration
SEI – Stockholm Environment Institute
SIPRI – Stockholm International Peace Research Institute
TEEB – The Economics of Ecosystems and Biodiversity
The Earth Charter Initiative
The World Bank
UN/DESA – United Nations Department of Economic and Social Affairs
UNDP – United Nations Development Programme
UNEP – United Nations Environment Programme
UNESCO – United Nations Educational, Scientific and Cultural Organization
U.S. Census Bureau
US/EIA – United States Energy Information Administration
Waterfootprint Network
Worldwatch Institute
World Wildlife Fund
WRI – World Resources Institute

Notes

Introduction

1. Andrade, Carlos Drummond de: "Ficar em casa". Jornal do Brasil, April 3, 1960. In Uma Pedra no Meio do Caminho. Biografia de um Poema. São Paulo: Instituto Moreira Salles, 2010.

Preface

1. State Decree No. 9.452 of December 5, 1982.
2. Institut d'Études Politiques de Paris. www.sciencespo.fr/
3. Pessoa, Fernando: A Little Larger Than The Entire Universe: Selected Poems. Edited and translated by Richard Zenith. New York: Penguin Books, 2006.
4. Eliot, T. S. "Burnt Norton" in Poesia. Translation, editing and notes by Ivan Junqueira. Rio de Janeiro: Nova Fronteira, 1981. Eliot, T. S. "Burnt Norton" In Four Quartets. New York: Harcourt Brace Jovanovich, 1943.

Dialogue in two periods

1. Paris Exposition of 1855. Chandler, Arthur: "Fanfare for the New Empire. The Paris Exposition of 1855". World's Fair magazine, vol. VI, No. 2, 1986. charon.sfsu.edu/publications/PARISEXPOSITIONS/1855EXPO.html

Chapter 1

1. Klabin, Israel: "Compreendendo a Cultura no Desenvolvimento Sustentável: investindo em legados culturais e naturais", World Bank and

UNESCO, September 28-29, 1998, World Bank Annual Meeting. Israel Klabin's collection.

2. Klabin, Israel: "Desenvolvimento Sustentável – Harmonia entre Ecologia e Economia." Fórum da Sociedade Civil para o Meio Ambiente ACRJ/Light 1994 – Painel sobre Cultura e Meio Ambiente Relações com o Desenvolvimento Econômico. Israel Klabin's collection.
3. Donella H. Meadows; Dennis L. Meadows; Jorgen Randers; William W. Behrens III: The Limits to Growth. A report for the Club of Rome's Project on the Predicament of Mankind. Washington, DC: Potomac Associates, 1972. (p. 93-94 e p. 197-198)
4. Drexhage, John; Murphy, Deborah: Sustainable Development: from Brundtland to Rio 2012. New York: UN Headquarters, September 2010. (p. 2)
5. United Nations Conference for the Environment and Development: Agenda 21. Rio de Janeiro, United Nations, 1992.
6. Klabin, Israel: "A sociedade civil e o setor privado no desenvolvimento sustentável." Rio de Janeiro: SID – Society for International Development, May 27, 1992 (Marina Palace Hotel). Israel Klabin's collection.
7. Strong, Maurice: Where on Earth are We Going? Toronto: Random House of Canada, 2000. ("Introduction," p. 1)
8. Elkington, John: "Enter the Triple Bottom Line" IN Henriques, Adrian; Richardson, Julie: The Triple Bottom Line, Does it All Add Up? Assessing the Sustainability of Business and CSR. London: Earthscan, 2004.
9. Klabin, Israel: Terra: Limites da Sustentabilidade. Rio de Janeiro: Eva Klabin Rapaport Foundation, 2001. (lecture given on November 29). Israel Klabin's collection.
10. Putnam, Robert: Bowling Alone: America's Declining Social Capital. Journal of Democracy 6:1, Jan, 65-78, 1995.
11. Klabin, Israel: "Inaugural speech." Official opening ceremony of Rio+5 Forum. Israel Klabin's collection.
12. Strong, Maurice: Where on Earth are We Going? Toronto: Random House of Canada, 2000. ("Rio +5: Successes and Failures," p. 282)
13. The Earth Charter Initiative: Carta da Terra. Paris: UNESCO, 2000. http://www.cartadaterrabrasil.org/prt/text.htmlandhttp://www.earthcharterinaction.org/content/pages/Read-the-Charter.html

14. United Nations Conference on Environment and Development: Convention on Biological Diversity. Rio de Janeiro: United Nations, 1992. http://www.onu-brasil.org.br/doc_cdb1.php and http://www.cbd.int/convention/text/
15. Bovarnick, A.; F. Alpizar; C. Schnell, Editors. The Importance of Biodiversity and Ecosystems in Economic Growth and Equity in Latin America and the Caribbean: An economic valuation of ecosystems, United Nations Development Programme, 2010.
16. Klabin, Israel: "Remarks". OECD: Paris, January 25, 2001. International Workshop on Market Creation for Biodiversity Products and Services. Israel Klabin's collection.
17. Walker, Brian; Salt, David: Resilience Thinking. Sustaining Ecosystems and People in a Changing World. Washington: Island Press, 2006.
18. Diamond, Jared: Colapso. Como as sociedades escolhem o fracasso ou o sucesso. Rio de Janeiro: Record, 2005. Excerpts from p. 582 e p. 584. Diamond, Jared: Collapse. How societies choose to fail or succeed. New York: Viking, 2005. Excerpts from Chapter 16.
19. UN/DESA Population Division: World Population to 2300. New York: United Nations Department of Economic and Social Affairs: 2004.
20. UN/DESA Population Division: 2010 Revision of World Population Prospects Press Release. New York, United Nations Department of Economic and Social Affairs, 2011.
21. U.S. Census Bureau: U.S. and World Population Clocks. Washington: U.S. Census Bureau, 2011. www.census.gov/main/www/popclock.html
22. Footprint Network. www.footprintnetwork.org
23. Waterfootprint Network. www.waterfootprint.org
24. Klabin, Israel: "Que tipo de mundo será o século XXI?" Lecture given at AMATRA, May 9, 2002. Israel Klabin's collection.
25. World Wildlife Fund: Living Planet Report 2010. Biodiversity, biocapacity and development. Gland (Switzerland): WWF, 2010.
26. Klabin, Israel and Salati, Eneas: "Algumas Observações Sobre a Avaliação dos Impactos Ambientais e Hidrológicos do Projeto da Hidrovia Paraguai-Paraná" In O Projeto de Navegação da Hidrovia Paraguai-Paraná. Relatório de uma Análise Independente. Brasília: EDF/CEBRAC, July 1997. Israel Klabin's collection.

27. Klabin, Israel: "A controvertida hidrovia Paraguai-Paraná". São Paulo: Folha de São Paulo, March 14, 1997. Israel Klabin's collection.
28. Spalding M.D.; Ravilious C.; Green E.P.: World Atlas of Coral Reefs. Berkeley: University of California Press, 2001. UNEP – Coral Atlas. UNEP – WCMC Coral Reef Unit.
29. Worldwatch Institute: State of the World 2010. Transforming Cultures. From Consumerism to Sustainability. London: W.W. Norton & Company, 2010.
30. Bíblia Hebraica. São Paulo: Editora e Livraria Sêfer Ltda., 2006. Hebrew Bible. A Hebrew – English Bible According to the Masoretic Text and the JPS 1917 Edition. 2005 http://www.mechon-mamre.org/p/pt/pt0.htm

Chapter 2

1. Becker, Bertha; Stenner, Claudio: Um futuro para a Amazônia. São Paulo: Oficina de Textos, 2008. (Série Inventando o Futuro.)
2. Salati, Eneas: "Modificações da Amazônia nos últimos 300 anos: suas consequências sociais e ecológicas." IN Brito, Sergio: Desafio amazônico: o futuro da civilização dos trópicos. Brasília: UnB/CNPq, 1990.
3. Klabin, Israel; Marques, Maria Silvia Bastos; Ricupero, Rubens; Reichstul, Phillipe: "O Futuro da Amazônia é o nosso futuro." São Paulo: Folha de São Paulo, 2005. Israel Klabin's collection.
4. Gesisky, Jaime: "Secas severas na Amazônia deixam cientistas em alerta". Altamira: IPAM, February 4, 2011. http://www.ipam.org.br/noticias/
5. Salati, Eneas, interview in the promotional video for Rios Voadores project – www.riosvoadores.com.br. Rio de Janeiro: Safari Air Empreendimentos Ltda, December 2010.
6. Klabin, Israel: "Uma agenda para florestas tropicais". São Paulo: Folha de São Paulo, June 26, 1997.
7. Instituto Brasileiro de Geografia e Estatística: Produção da Pecuária Municipal 2007. Rio de Janeiro: IBGE, November 26, 2008.
8. Andrade, Carlos Drummond de: "Adeus a Sete Quedas". Rio de Janeiro: Jornal do Brasil, September 9, 1982.

Chapter 3

1. Klabin, Israel: lecture: FAO Advisory Committee on Paper and Wood Products, Fortieth session. São Paulo: FAO, April 28, 1999. Israel Klabin's collection.
2. Cook, Earl: "The flow of energy in an industrial society", in Man, Energy and Society. W. H. Freeman Edition, 1976.
3. Earth at Night: C. Mayhew & R. Simmon (NASA/GSFC), NOAA/NGDC, DMSP Digital Archive, October 5, 2008.
4. Klabin, Israel: lecture: FAO Advisory Commitee on Paper and Wood Products, Fortieth session. São Paulo: FAO, April 28, 1999. Israel Klabin's collection.
5. Bergreen, Lawrence: Marco Polo, de Veneza a Xanadu . Rio de Janeiro: Editora Objetiva, p. 168. The travels of Marco Polo, the Venetian: the translation of Marsden revised, with a selection of his notes. London: H. Bohn, 1854. (p. 229-230)
6. Marinetti, Filippo: Manifesto of Futurism. Translated by James Joll. Paris: Le Figaro, February 20, 1909.
7. Adam Nossiter, "Nigéria enfrenta vazamento de petróleo com cinco décadas de existência". The New York Times, trans. UOL. June 17, 2010.
8. Frank Dohmen et al., "Demanda mundial de Carvão é cada vez maior." Der Spiegel (Germany), trans. UOL: July 25, 2010.
9. ibid.
10. U.S. Energy Information Administration. International Energy Outlook 2010. Washington DC: U.S. Department of Energy, 2010.
11. ibid.
12. International Energy Agency / OECD: "Clean energy progress report." IEA / OECD, 2011.
13. Friedman, Thomas: "Who's Sleeping Now?" New York: The New York Times, February 9, 2010.
14. Jornal Valor Econômico, "Editorial", June 6, 2011.
15. Thomas Friedmann, The New York Times, "Crise no meio ambiente vai obrigar pessoas a consumirem menos", June 8, 2011, trans. UOL. http://www.bbc.co.uk/news/world-asia-pacific-12595872
16. Klabin, Israel: "Energia renovável e eficiência energética para o desenvolvimento sustentável". Palestra Seminário REIA, July 9, 1997. Israel Klabin's collection.

17. World Bank Poverty Reduction and Equity group: Food Price Watch. World Bank: April 2011.
18. The scope of fossil-fuel subsidies in 2009 and a roadmap for phasing out fossil-fuel subsidies. An IEA, OECD and World Bank Joint Report prepared for the G-20 Summit, Seoul (Republic of Korea), November 11-12, 2010.
19. Klabin, Israel: "Conclusão" IN Schindler, Walfredo (org.): 1º Seminário Brasileiro sobre CCS – Carbon Dioxide Capture and Storage. Um debate sobre os desafios dessa nova tecnologia. Rio de Janeiro: Brazilian Foundation for Sustainable Development – FBDS/Shell Brazil, 2009.
20. International Energy Agency/OECD: "Clean energy progress report". IEA/OECD, 2011.

Chapter 4

1. Berlin, Isaiah: Estudos Sobre a Humanidade. Uma Antologia de Ensaios [The Proper Study of Mankind: An Anthology of Essays]. Rio de Janeiro: Cia. Das Letras, 2002.
2. Luzes do Império. D. Pedro II e o Mundo Judaico. Sesc São Paulo/Casa de Cultura de Israel, 1999.
3. ERNEST, Renan. Qu'est-ce qu'une nation? Conference at the Sorbonne, March 11, 1882. (p. 10-11) ERNEST, Renan. "What is a Nation?" IN Eley, Geoff and Suny, Ronald Grigor, ed. 1996. Becoming National: A Reader. New York and Oxford: Oxford University Press, 1996.
4. Kant, Emmanuel: "Ideia de uma História Geral com um Propósito Cosmopolita."
5. Klabin, Israel: "State, Peace and Power: an exercise on democracy (the three basic rights)", Bombay, India, 1990. Israel Klabin's collection.
6. World Bank: Governance and Development. Washington, DC: World Bank, 1992.
7. Klabin, Israel: "Marte e a destruição da Amazônia". Rio de Janeiro: O Globo, July 10, 1997.
8. Cardoso, Fernando Henrique: "Global Governance in the XXI Century". Series of debates about the international political and economic Outlook. London: London School of Economics and Political Science (LSE), November 16, 2004.

9. Cardoso, Fernando Henrique: "The need for Global Democratic Governance: a perspective from Latin America". Washington, DC: Library of Congress, February 22, 2005. "Henry Kissinger Lectures in Foreign Policy and International Relations".
10. Israel, Klabin: "Carta a Fernando Henrique Cardoso." Rio de Janeiro, September 17, 1997. Israel Klabin's collection.
11. Cardoso, Fernando Henrique: "Carta ao Senhor Israel Klabin". Brasília, October 23, 1997. Israel Klabin's collection.
12. Lampedusa, Giuseppe Tomasi di: Il Gattopardo. Milano: Le Comete / Feltrinelli, 2002.
13. Brown, Lester: Plano B 4.0. Mobilização para salvar a civilização [Plan B 4.0 – Mobilizing to Save Civilization]. São Paulo: New Content, 2009.
14. Editorial from the newspaper O Estado de São Paulo, August 3, 2010.
15. Hobsbawm, Eric: Era dos Extremos, São Paulo: Companhia das Letras, 1995. (p. 56). Hobsbawm, Eric: Age of Extremes, London: Abacus, 1995. (p. 49).
16. Toynbee, Arnold: Um Estudo da História. Rio de Janeiro: Martins Fontes, 1986. (p. 256-257.)
17. ibid.

Chapter 5

1. Calvino, Italo: As Cidades Invisíveis. São Paulo: Companhia das Letras, 2000. Calvino, Italo: Invisible Cities. Orlando: Harcourt Trade Publishers, 1974.
2. Op. Cit. Gudin, Eugênio: "Israel Klabin no Leme do Burgo Podre." Rio de Janeiro: O Globo, 1979. Israel Klabin's collection.
3. The Ministério da Desburocratização [Ministry of Debureaucratization] existed between 1979 and 1986 in an attempt to reduce the costs of bureaucratization on the economy and social life in Brazil.
4. Law 105, from July 13, 1979.
5. Klabin, Israel: speech on September 27, 1979. Israel Klabin's collection.
6. "Aprendendo com o patrimônio". IN Oliveira, Lúcia Lippi (org.) Cidade: História e Desafios. Rio de Janeiro: Editora FGV, 2002.

7. Research on documents about Israel Klabin's administration of the city of Rio de Janeiro, by Paulo Rubens Sampaio.
8. Publication from the city of Rio de Janeiro/Secretaria Municipal de Planejamento e Coordenação Geral. Israel Klabin's collection.

Chapter 6

1. Campos, Roberto: "Bretton Woods, FMI, BIRD, Havana e GATT: A Procura da Ordem Econômica do Após-Guerra". Brasília: Boletim de Diplomacia Econômica – MRE, February 19, 1995.
2. Keynes, John Maynard, «National Self-Sufficiency,» The Yale Review, Vol. 22, nº 4 (June 1933), p. 755-769.
3. Daly, Herman: "Sustainable growth? No thank you" IN Mander, Jerry & Goldsmith, Edward (eds.): The Case Against the Global Economy and for a turn toward the local. San Francisco: Sierra Club Books, 1996, p. 192-96.
4. Judt, Tony: O Mal Ronda a Terra. Um tratado sobre as insatisfações do presente [Ill Fares the Land]. Rio de Janeiro: Editora Objetiva, 2011.
5. Klabin, Israel: "The Liberal Issues and The Third World". Rio de Janeiro, September 1966.
6. Myers, Norman: Ultimate Security: the Environmental Basis of Political Stability. New York: W.W. Norton & Co., 1993. (p. 222.)
7. Brown, Lester: Plano B 4.0. Mobilização para salvar a civilização [Plan B 4.0 – Mobilizing to Save Civilization]. São Paulo: New Content, 2009.
8. Kennedy, Robert: Remarks at the University of Kansas. Kansas: March 18, 1968. www.jfklibrary.org
9. Friedlingstein et al. Update on CO_2 Emissions. Nature Geoscience, doi:10.1038/ngeo_1022. Published on-line, November 21, 2010. Global Carbon Project: Carbon Budget and Trends 2009. www.globalcarbonproject.org/carbonbudget
10. Elkington, John: Cannibals with Forks. The Triple Bottom Line of 21st Century Business. Oxford: Capstone Publishing, 1997. (Foreword, vii.)
11. Klabin, Israel: "Financiamento para Projetos de Meio Ambiente – dificuldades e oportunidades". Workshop Meio Ambiente – Oportunidades de Negócios e Riscos a Curto e Longo Prazo – November 12, 1996. Israel Klabin's collection.

12. Klabin, Israel: "Prefácio à edição brasileira" IN Stern, Nicholas: O Caminho Para um Mundo Mais Sustentável. Rio de Janeiro: Campus-Elsevier, 2010.
13. Klabin, Israel: "Preservando o Futuro". Rio de Janeiro: Jornal do Brasil, June 1, 1992. Israel Klabin's collection.
14. Marcovitch, Jacques (editor); Margulis, Sergio and Dubeux, Carolina Burle Schmidt: Economia da Mudança do Clima no Brasil: Custos e Oportunidades. São Paulo: IBEP Gráfica, 2010.
15. Such as Academia Brasileira de Ciências (ABC), Instituto de Pesquisa Econômica Aplicada (IPEA), Faculdade de Economia Administração e Contabilidade da Universidade de São Paulo (FEA/USP), Instituto Nacional de Pesquisas Espaciais (INPE) e a Fundação Brasileira para o Desenvolvimento Sustentável (FBDS). [Brazilian Academy of Sciences (ABC), Institute of Applied Economic Research (IPEA), of Economics, Business and Accounting of the University of São Paulo (FEA/USP); Institute for Space Research (INPE) and Foundation for Sustainable Development, (FBDS).]
16. Marcovitch, Jacques (editor); Margulis, Sergio and Dubeux, Carolina Burle Schmidt: Economia da Mudança do Clima no Brasil: Custos e Oportunidades. São Paulo: IBEP Gráfica, 2010. (p. 75)
17. Brazilian Foundation for Sustainable Development (FBDS): "Estimativas da oferta de recursos hídricos no Brasil em cenários futuros de clima (2015-2100)." Commented by Eneas Salati (editor), Walfredo Schindler, Daniel de Castro Victoria, Eneida Salati, João Carlos Simanke de Souza and Nilson Augusto Villa Nova.
18. Stern, Nicholas: "A conta ficou mais alta – entrevista a Ana Luiza Herzog". São Paulo: revista EXAME, August 11, 2010. 18. www.walmartbrasil.com.br/arquivo/1180-A-wm_sustentabilidade_clima.pdf and http://planetasustentavel.abril.com.br/noticia/desenvolvimento/entrevista-economista-nicholas-stern-catastrofe-ambiental-exame-586961.shtmlf
19. TEEB for National and International Policy Makers. Summary: Responding to the Value of Nature. TEEB – The Economics of Ecosystems and Biodiversity (2009), http://teebweb.org/
20. Gallai, N.; Salles, J-M.; Setelle, J e Vaissière, B.E.: Economic Valuation of the Vulnerability of World Agriculture Confronted with Pollinator Decline. Ecological Economics (2009), Vol. 68 (3): 810-21.

21. UNEP, 2011, Towards a Green Economy: Pathways to Sustainable Development and Poverty Eradication – A Synthesis for Policy Makers, www.unep.org/greeneconomy
22. Smith, Adam: The Theory of Moral Sentiments. London: A. Miller, 1790, 6th edition. (VI.II.46) http://www.econlib.org/library/Smith/smMS.html

Chapter 7

1. The data presented in this chapter is from the Global Carbon Project: Carbon Budget and Trends 2009. (annual CO_2 emissions) www.globalcarbonproject.org/carbonbudget and from NOAA, National Oceanic & Atmospheric Administration, www.esrl.noaa.gov, U.S. Department of Commerce (average stocks of CO_2 in the atmosphere).
2. Klabin, Israel and Ricupero, Rubens: "A Conferência de Bali e suas consequências para a humanidade". Rio de Janeiro: O Globo. November 28, 2007.
3. Vernant, Jean-Pierre: As Origens do Pensamento Grego. Rio de Janeiro: Bertrand Brasil, 2002. Vernant, Jean-Pierre: The Origins of Greek Thought. Ithaca: Cornell University Press, 1982.
4. As proposed by physicist Stephen Hawking in Hawking, Stephen: Os Gênios da Ciência. Sobre os Ombros de Gigantes. As mais importantes ideias e descobertas da física e da astronomia organizadas e comentadas pelo mais famoso físico da atualidade. Rio de Janeiro: Editora Campus-Elsevier, 2005.
5. ibid.